REVISE
BIOLOGY

A COMPLETE REVISION COURSE FOR
GCSE

Julian Ford-Robertson   MA(Oxon)
Head of Biology Department, Haileybury College, Hertford

Charles Letts & Co Limited
London, Edinburgh & New York

First published 1979
by Charles Letts & Co Ltd
Diary House, Borough Road, London SE1 1DW
Revised 1981, 1987

Illustrations: Stan Martin, Ann Savage and Chartwell Illustrations

British Library Cataloguing in Publication Data
Ford-Robertson, Julian
Revise biology : a complete revision course for GCSE. — 5th ed — (Letts study aids)
1. Biology
I. Title
574   QH308.7
ISBN 0 85097 775 4

'Letts' is a registered trademark of Charles Letts & Co Ltd

Printed and bound in Great Britain by Charles Letts (Scotland) Limited

PREFACE

This new edition of *Revise Biology* is designed to meet the needs of the new GCSE and SCE exams. It is not simply a new cover around the original book which has served countless pupils sitting GCE, CSE and SCE so well for many years. It has been completely revised and reshaped to suit new needs. It emphasizes the **experimental basis** for scientific knowledge, putting the facts deduced by experiment into a **framework of knowledge**. Advice on **revision** and **examination technique** is given along with **questions** on which students may gain practice. In the last unit **ideas for experiments of your own** are given.

In the preparation of this book I have been greatly assisted by helpful criticism from C. G. Gayford, BSc, MEd (Science Education), PhD, PGCE, Lecturer in Education at the University of Reading; Wilf Stout, BSc, MA, MEd, CBiol, MIBiol; and B. Arnold MSc, MIBiol, Lecturer in Biology at Aberdeen College of Education. I am also grateful for specialist advice from Mr P. H. L. Worth, Consultant Urologist, Dr J. Higgo of Imperial College and Mrs E. Skinner of the Cancer Research Campaign. To my wife and Susan Hunt who typed, to Stan Martin who illustrated and to Mrs Eileen Lloyd who edited the work I owe a special debt; but numerous others not named — family, colleagues and friends — also lent valuable support. The patience and understanding of the staff at Charles Letts & Co. Ltd, in particular Ms Diane Biston and Mrs Julia Cousins, was also appreciated.

I am grateful to the following Examination Groups for permission to use a selection of their GCSE and Standard Grade (SEB) specimen questions in Biology: LEAG, MEG, NEA, NISEC, SEG, SEB.

Julian Ford-Robertson
1987

CONTENTS

INTRODUCTION

How to Use This Book

Revise Biology is written especially for those who need help in preparing for the GCSE or the Scottish Ordinary or Standard Grade examinations. It provides:

— advice on **what your syllabus requires**: a table of syllabus analysis.
— advice on **how to learn**: learning made easier.
— **what to learn** in readily revisable form: information, lavishly illustrated.
— advice on how to **show the examiner** that you know what he is asking for: an outline of good examination technique.
— **practice in answering** examination questions.

If you follow this sequence in the use of this book, you will have a good chance of success.

USING THE TABLE OF ANALYSIS

Turn to p. xiv and select from the top of the page your own Examination Group syllabus. The key to the initials for the Groups may be found on p. xviii. Below the name of the syllabus are details of:

(*a*) the number of theory papers and their length;
(*b*) the percentage of the total mark awarded for your Coursework Assessment;
(*c*) what you need to know, topic by topic. This material has been divided into numbered units, not all of which need to be studied for your own syllabus.

Select carefully the units you require by referring to the symbols:

- • unit required for a syllabus
- (•) unit required, but with reservations. These reservations may include such things as 'less detail than this is required' or 'this organism can be used as an example for a part of the syllabus, but is not specifically named'.

Where neither of these symbols appears against a topic, i.e. the space is *left blank*, you do not need to study that unit.

The table of analysis must be regarded only as a helpful guide. If you send for your Examination Group's syllabus and for copies of past examination papers (use the addresses on p. xviii) you will be able to judge more easily what the special features of your syllabus are. Your teacher will also give you advice, particularly on units marked (•).

Before using the subject material, first take care to understand *how* you should revise (see 'Success in Exams' below). Do not exceed your 'concentration time' (see p. xi).

USING THE SUBJECT MATERIAL

Work through only the units that you need to select (see above). Use every memory aid that you can (see 'Success in Exams'). Remember that 70–80% of the marks that decide your grade in GCSE (or equivalent) can come from **your efforts on paper in the examination room**.

The other 20–30% of your marks can come from practical work **in the laboratory** (Course Assessment). So remind yourself of the practical techniques that you should have used throughout the course. Unit 22 is particularly important.

EXAMINATION TECHNIQUE FOR BIOLOGY EXAMINATIONS

Turn to p. 165. The advice given ranges from tips on organizing what to take into the examination room to how to use your time well.

PRACTICE IN ANSWERING EXAMINATION QUESTIONS

Turn to p. 169. Build up your confidence by trying to answer these examination questions — all of them provided by the Examining Groups.

Success in Exams

Successful students are those who can organize their work. In particular, they must be able to work effectively on their own. If you are to be successful you need determination to succeed, a work plan fitted to a time schedule and determination to keep to that schedule. Unfortunately, few students are told *how* to devise that plan and carry it out — that is where this book comes in.

This section contains some advice that will help you to succeed in school. It also gives reasons for this advice. The rest of the book concerns itself with presenting biological facts, and how to deduce them by experiment, in a form that makes it easy to revise.

The first three steps in the learning process are planned by your teacher who knows the sort of examination you will be sitting (stage 5 in Fig. A) and plans accordingly. Where so many students fail, needlessly, is at stage 4 (revision) — because they do not know how to go about it. **Revision** is what this book is all about — leave out stage 4 in the diagram below and you have F for failure.

The S that signals success · · ·

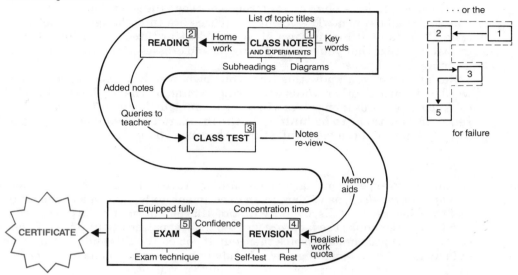

Fig. A The 'S' for success ... or the 'F' for failure

THE LEARNING PROCESS: PATTERNS IN THE MIND

In science you learn from experiments — your own or those reported by others. It is well known that students tend to remember far better the 'facts' they have learned by doing experiments themselves. Unfortunately there is not enough time to learn everything this way, so that the rest has to be learned by reading and listening.

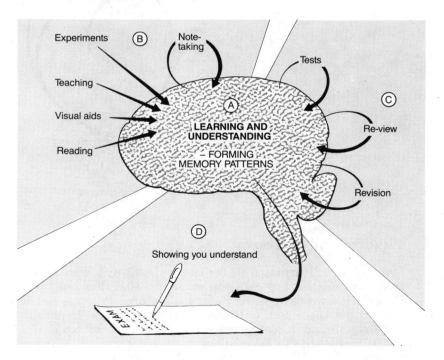

Fig. B From learning to showing you understand

Why is learning through reading and listening harder than learning by experiment? Why are good annotated diagrams often so much easier to learn than line upon line of words? Why is a good teacher such a help in learning?

Experiments

Doing these needs personal involvement and the use of several senses. Then, at the end of the experiment, one must arrive at a conclusion — which requires some reasoning. In a word, the whole process requires *understanding* — understanding the **aim** of the experiment, the **method** to be used and how to record the **results** in a meaningful way. And the final step, the **conclusion,** requires reasoning from what you have already understood.

During this process you will notice that you have built up a pattern of knowledge — like a jig-saw — lacking just one piece to complete it (the conclusion). Having completed it, if you were a professional scientist, you would find yourself doing yet another jig-saw, and another, and so on. However, there is one vital difference between the scientist's jig-saw game and the jig-saws you do for amusement. The scientist's pieces are interchangeable between *different* jig-saws (Fig. C). It is as if his jig-saws all interlocked in three dimensions. On a simpler level you probably realize that you can play noughts and crosses in three dimensions and not just in two, as on a piece of paper.

But of course you know this from your own everyday experience. When someone says the word 'cat' various other words will spring to mind. Perhaps 'claws', 'fluffy', 'leopard' or 'witches', and these words themselves will bring to mind further words. In short, **learning is a process requiring patterns to be built up in the mind:** relate what you have just learned to what you already know and the facts will stick — because you *understand*.

Reading

In contrast to experimenting, when reading you are using only one of your senses — sight — and you are not *involved*, as you are in an experiment, unless you make a mental effort. Nor do you feel the same sense of discovery. Worst of all, the information is presented as a series of facts. In well-written books the facts *are* written to form patterns; but it is you, the reader, who has to concentrate hard enough to pick them out. If you can do this and fit your newly acquired jig-saw of facts into the system of jig-saws already in your brain, you have learned to learn by reading. This is initially much harder to do than by experiment.

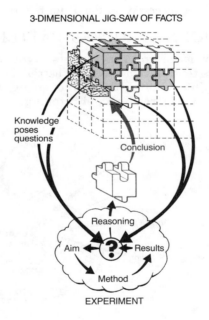

3-DIMENSIONAL JIG-SAW OF FACTS

Knowledge poses questions

Conclusion

Reasoning

Aim — Results

Method

EXPERIMENT

Fig. C Learning

Pictures and diagrams

One method of learning from books goes half-way towards experimental learning. Do you enjoy strip cartoons? At any rate you will agree that they are easy reading and convey much more than the few words appearing with the pictures. Pictures and diagrams, like words, require eyes alone to see them. But, unlike words, pictures build up patterns in the brain more readily and understanding is more immediate. So, well-constructed diagrams are an invaluable learning aid. If you learn the art of diagram-drawing you will reinforce both your memory and your understanding. Ultimately you should be able to construct your own original topic-summary diagrams, and there is a definite stage in the learning process when you should do this (see 'Retaining facts').

Teachers

Have you ever thought about the role of teachers in the learning process? They attempt to activate more senses in you than just your hearing. By showing films and slides, by drawing diagrams or asking your opinions and by giving you definite learning objectives, they try to keep you personally involved. It is for you to respond — if you are going to learn. Amidst it all you must carefully latch onto the *pattern* of facts that the teacher explains. A teacher usually explains what the *whole* lesson is to be about during the first few minutes. Listen hard to that outline and the rest of the lesson will be easier to absorb. The outline is the basic skeleton upon which the teacher will build up the flesh and features of the subject, as the lesson proceeds. If you miss the description of the skeleton, the subject may turn out to be a monster for you!

CAPTURING FACTS

Class notes

Your teacher has probably advised you on how to make these. For easy revision it is essential that they include:

(*a*) a topic list referring to numbered pages in your notebook;
(*b*) clear, underlined topic titles and sub-titles;
(*c*) underlined 'key words';
(*d*) clear diagrams with titles;
(*e*) space for topic-summary diagrams made during revision.

Reading texts

Your teacher may advise you what to read. Realize too that a text has an index at the back; use it to look up things for yourself. At this stage many students get bogged down because they read slowly and give up. If you are someone with this problem, try this:

The cat sat on the mat.

Because of the way you were taught to read, for example 'c-a-t' or 'cat', you have been 'brainwashed' into thinking that you can only read one word at a time. Now bring your head back further from the page. Notice that now you can have more than one word in focus at a time — without having to move your eyes at all. With practice you will find that not only 'cat' is in focus but also 'The' and perhaps even 'sat' as well. It does need practice but soon you will find that the whole of 'The cat sat on' is in focus at one glance and that you can take it *all* in. Four words instead of one at each glance — four times your original reading speed!

Time how long it takes you to read a page now. Repeat the test after each week of practising the new method. Some people can read 800 words per minute with ease, understanding as they go. No wonder this method is called speed-reading! Reading the text should be done after you have been taught the topic — say during home-work. Your reading:

(*a*) reinforces in your mind the facts recorded in class notes;
(*b*) allows you to add extra bits to your notes;
(*c*) should clear up misunderstandings.

Ask your teacher if you still do not understand something.

Class tests

These are designed to help you to recall facts and to reason from them. In this process you and the teacher are on the *same* side; together you will succeed. The teacher is *not* putting you to the torture. Tests:

(*a*) help you to assess your progress (should you work harder?);
(*b*) help the teacher to clear up your difficulties (adjust your notes?);
(*c*) help you to remember facts better;
(*d*) give you exam practice.

RETAINING FACTS

Revision

This is the vital last stage in the learning process, the stage when you are finally on your own. You must understand clearly how to go about it. Look at the graph in Fig. D.

All of us have different **'concentration times'**. How long is yours? Go to a quiet working place indoors, without distractions, and note the time. Read a part of your text-book that is new to you, making a determined effort to take in all you read. When your mind begins to wander, look again at your watch; you are at the end of your concentration time. It should be around 20–40 minutes and will differ according to the amount of sleep you have had, what else is on your mind, and even on the subject matter. Never revise for longer than your concentration time. If you do, you will waste your time. You may still be reading but you will not understand. So **rest** for five minutes.

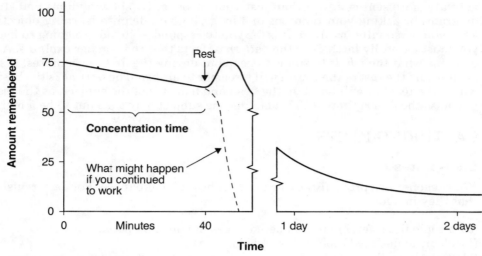

Fig. D Concentration time graph

After the rest, surprisingly enough, the facts you read in the text-book will come back to you more easily still. During the rest, the brain was 'organizing' the facts you took in. Note-taking would have assisted this organizing process. Unfortunately most of these facts go into what is called your 'short-term memory'. Within 48 hours you will retain as little as 10% of what you thought you knew so well. Don't be depressed. You can push these facts into your 'long-term memory', which is essential for examination purposes, by **reviewing**.

Reviewing

This is a *quick* re-read of your notes, taking only a few minutes. If your notes are disorganized you will not gain much. But with clear summaries, such as you will find in this book, you should dramatically increase the number of facts going into your long-term memory. Do this re-reading after a week and then again two weeks later after having learned the topic for the first time in class. Figure E shows the sort of result that can be obtained by thorough reviewing and revision.

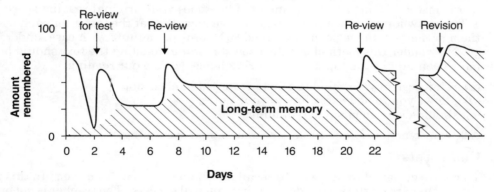

Fig. E Review and revision graph

Revision is just an extension of reviewing. If you have followed the learning plan so far, there will be relatively little to do. During revision whole chunks of your notes will not need to be read because sub-titles and key words alone will trigger off a mass of facts already in your long-term memory. For the rest of the plan, follow these principles:

(i) Months ahead of the examinations plan how much to revise each week.
(ii) Have a regular time for work and stick to it.
(iii) With your concentration time in mind, plan a *realistic* amount of work for each 20–40 minute session. You must get up from your task with a sense of achievement, i.e. that you have completed what you set out to do. Otherwise you will get depressed 'at the hopelessness of it all'.
(iv) Take those 5-minute breaks. But do not exceed them.
(v) Use the memory aids and summary diagrams in this book to help you.

Memory aids

(i) Repetition   (ii) Mnemonics   (iii) Pattern-diagrams

(i) **Repetition:** By chanting something over and over again you can learn it 'parrot-fashion'. Many people learn their times-tables or poetry in this way. The method has its uses. But though you can remember in this way you do not necessarily *understand*.

(ii) **Mnemonics:** These are words, sentences or little rhymes chosen from everyday language to help you to remember technical words that you find difficult to memorize. This book provides you with a few examples; but you may be able to do better. Make your own mnemonics funny, outrageously absurd — even rude — if you are going to remember them. Dull mnemonics are difficult to remember. The words you choose must be sufficiently similar to the technical words to remind you of them. For example:

'How do I remember the words on the
Royal Garter when I don't know enough
French?' Look at Fig. F and try:

On his way he madly puns.

Fig. F The Royal Garter

'How can I remember the characteristics of living things — which I *do* understand but may not be able to remember fully in an exam?' Try **Germs in our seas** and turn to unit 1.1. This example uses initial letters of the key words only. You will find another mnemonic in unit 2.2.

(iii) **Pattern-diagrams:** these are important or 'key' words written down and joined up with lines according to their connections with each other. You have already seen two examples (Figs. A and B). When you have finished revising a topic always try to summarize it in this way. You will be surprised how easy it is. And why? Because, you will remember (Fig. C), your mind thinks in patterns and not in lists. When you come to the examination you will be able to remember your pattern-diagrams and even create new ones when planning your answers to essay questions.

Table of Analysis of Examination Syllabuses

GROUPS AND SYLLABUSES	LEAG A	LEAG B	LEAG C	MEG A	MEG B	NEA	NISEC	SEG	WJEC	SEB Standard Grade*	SEB Ordinary Grade	MEG Combined Science (Dual Award)
Papers and their length in hours for										(General)		(For 3 sciences)
1. Grades C and below	2 1½ OR	2 1½ OR	2 1½ OR	⅔ 1⅓ AND	⅔ 1⅓ AND	2 1½ OR	1 1¼ AND	2 — AND	2 — OR	1½ AND 1½ OR	OR	2 2 AND
2. Grades A and B	1½	1½	1½	1¼	1¼	1½	1¼	1½	2	1½ AND 1½ (Credit)	[1] AND [1½]	2
3. Marks (%) for Coursework Assessment	20	20	20	20	20	30	20	20	20		—	20
1 Life												
1.1 Characteristics of organisms	•	•	•	•	•	•	•	•	•	•	•	•
1.2 Cells in detail	(•)	(•)	(•)	•	•	(•)	•	(•)	(•)	•		•
1.3 How the nucleus controls the cell				(•)	•		•				•	•
1.4 Sizes of cells and their sub-units	(•)	(•)		(•)							(•)	
1.5 Enzymes and metabolism	•	•	•	•	•	•	•	•	•	•	•	•
1.6 Units of life beyond the cell	•	•	•			•	•	•				
2 Classification												
2.1 Linnaeus and his classification system				•				•		(•)	•	
2.2 Groups and sub-groups				•							•	
2.3 Plant kingdom	(•)	(•)	(•)	•	•	•	•	(•)	(•)		•	
2.4 Animal kingdom	(•)	(•)	(•)	•	•	•	•	(•)	(•)		•	
2.5 Modern classification												
2.6 Multicell green plants and animals compared								•				
3 Viruses, microorganisms, fungi and biotechnology												
3.1 Viruses						•		•			•	(•)
3.2 Bacteria	•	•	•	•	•	•	•	•	•	(•)	•	(•)
3.3 Importance of bacteria	•	•	•	•	•	•	•	•	•	•		•
3.4 How viruses and bacteria reach people	•	•	•	•	•	•	•	•	•	•	•	•
3.5 Control of harmful bacteria	(•)	(•)	•	(•)	(•)	(•)	•	•	•	•	•	
3.6 Growing bacteria	(•)	(•)	(•)	(•)	(•)		(•)	(•)	•	•		•
3.7 Fungi—moulds and their culture	•	•		•		•	(•)	•	•	•	•	
3.8 Fungi—mushrooms and yeasts	•					(•)	•	•	•			
3.9 Importance of fungi	•	•		(•)	•	•	•	•	(•)	(•)		
3.10 Algae	•	•			•	•	•	•	•		•	•
3.11 Protozoa					•		•	•	•		•	(•)
3.12 Biotechnology			•	•	•	(•)		•	(•)	•		(•)
4 Foods and feeding												
4.1 Food	•	•	•	•	•	•	•	•	•	•	•	•
4.2 Holophytic, holozoic and saprophytic nutrition compared	•	•	•	•	•		•	•	•		•	•
4.3 Mineral salts for mammals and angiosperms	(•)	(•)	(•)	(•)	(•)	(•)	(•)	(•)	(•)		•	(•)
4.4 Carbohydrates, fats and proteins	•	•	•	•	•	•	•	•	•		•	•
4.5 Vitamins	(•)	(•)	(•)	(•)	(•)	(•)	(•)	(•)	(•)		•	(•)
4.6 Diet	•	•	•	•	•	•	•	•	•		•	•
5 Green plant nutrition												
5.1 Photosynthesis	•	•	•	•	•	•	•	•	•	•	•	•
5.2 Factors necessary for photosynthesis	•	•	•	•	•	•	•	•	•	•	•	•
5.3 Limiting factors					•		•				•	
5.4 Rate of photosynthesis						(•)	(•)					•
5.5 Leaf structure and photosynthesis	•	•	•	•	•	•	•	•	•	•	•	•
5.6 Gaseous exchange in leaves	•	•	•	•	•	•	•	•	•	•	•	•
5.7 Amino-acid synthesis	•	•	(•)	•	(•)	(•)	(•)	•	•		(•)	•
5.8 Mineral salt uptake by roots	•	•		•			•	•				
6 Animal nutrition												
6.1 Feeding methods of animals			•		•					•	•	
6.2 Digestion and its consequences	•	•	•	•	•	•		•	•	•	•	•
6.3 Experiments with digestive enzymes	(•)	(•)	•	•	•		•	(•)	(•)	•	•	(•)
6.4 Mammal teeth			•	•	•		•	•	•	•	•	
6.5 Mammal alimentary canal			•	•	•		•	•	•	•	•	
6.6 Dental health	•	•										
6.7 Herbivores and carnivores: teeth and jaws			•		•					(•)	•	
6.8 Herbivores and carnivores: the gut											•	
6.9 Absorption of food at a villus	•	•	•			•	•	•	•	•	(•)	•
6.10 Storage of food	•	•	•	•			•	•	•	•	•	•

GROUPS AND SYLLABUSES

#		LEAG A	LEAG B	LEAG C	MEG A	MEG B	NEA	NISEC	SEG	WJEC	SEB Standard Grade*	SEB Ordinary Grade	MEG Combined Science (Dual Award)
	Animal nutrition (cont)												
6.11	The liver	(●)	(●)	●	(●)	(●)	●	●	●	●	(●)	●	(●)
7	**Water uptake and loss in plants and animals**												
7.1	Importance of water	●	●	●	●	●	(●)	●	●	●	●	●	●
7.2	Diffusion and active transport	●	●	●	●	●	●	●	●		●	●	●
7.3	Osmosis	●	●	●	●	●	●	●	●	●	●	●	●
7.4	Osmosis in cells	●	●	●	●	●	●	(●)	●	(●)	●	●	●
7.5	Water uptake and loss in flowering plants	●	●	●	●	●	●	●	●	●	●	●	●
7.6	Guard cells and stomata	(●)	(●)	●	●	●	●	●	●	●	(●)	●	●
7.7	Transpiration	●	●	●	●	●		●	(●)	●	●	●	●
7.8	Transport of organic food	●	●	●	●	●	●	●	●	●	●	(●)	●
7.9	Tissues in the stem and root	●	●	●	●	●	●	●	●	●			●
7.10	Water uptake and loss in animals					●							
8	**The blood and lymphatic systems**												
8.1	Blood systems	●	●	●	●	●		●	●	●	●	●	●
8.2	Mammal blood and other body fluids	●	●	●	●	●	●	●	●	●	●	●	●
8.3	Blood smears	(●)	(●)	●	(●)	●		(●)	(●)	●	●	●	
8.4	Blood vessels and blood circulation	●	●	●	●	●	●	●	(●)	●	●	●	●
8.5	The heart	●	●	●	●	●	●	●	(●)	●	●	●	●
8.6	Changes in blood around the circulatory system	●	●	●	●	●	●	●		●	●	●	●
8.7	Lymphatic system	●	●			●					●	●	
9	**Respiration**												
9.1	Breathing, gaseous exchange and cellular respiration	●	●	●	●	●	●	●	●	●	●	●	●
9.2	Cellular respiration (aerobic and anaerobic)	●	●	●	●	●	●	●	●	●	●	●	●
9.3	Anaerobic respiration			●	●	●	●	●	●	●	●	●	●
9.4	Aerobic respiration	●	●	●	●	●	●	●	●	●	●	●	●
9.5	Rate of respiration	(●)	(●)	(●)					●		(●)		
9.6	Gaseous exchange	●	●	●	●	●	●	●	●	●	●	●	●
9.7	Organisms respiring in water and air	(●)				●						●	
9.8	Mammal respiration	●	●	●		●	●	●	●	●	●	●	●
9.9	Gas changes during breathing	●	●	●		●	●	●	●	●	●	●	●
9.10	The respiratory pathway	●	●	●	●	●	(●)	●	●	●	●		●
9.11	Smoking and health	●	●	●	●	●		●	●	●			●
9.12	Gaseous exchange in flowering plants	●	●	●	●	●	●	●	●	●		●	●
9.13	Uses of energy from respiration			●	●	●	●		●	●	●	●	
9.14	ATP (adenosine triphosphate)			●	●	●							
9.15	Measuring energy values of foods						●		●	●			
10	**Excretion, temperature regulation and homeostasis**												
10.1	Wastes and means of excretion	●	●	●	●	●	●	●	●	●	●	●	●
10.2	Mammal urinary system	●	●	●	●	●	●	●	●	●	●	●	●
10.3	The nephron	●	●	●		●		●	●	●	●	●	●
10.4	Water conservation			●	●	●	●	(●)	●	●	●		●
10.5	Abnormal kidney function				●		●		●	●	●		●
10.6	Body temperature in organisms	(●)	(●)	(●)	(●)	(●)	(●)	(●)	(●)		(●)		(●)
10.7	Mammal temperature control	●	●	●	●	●	●	●	●		●		●
10.8	Temperature control in other organisms	(●)	(●)	(●)	(●)	(●)	(●)	(●)	(●)		(●)		(●)
10.9	Homeostasis	(●)	(●)	●	●	●	●	(●)	●		●		●
10.10	Skin functions	●	●		●	●	(●)	●	●		(●)		(●)
11	**Sensitivity**												
11.1	Sensitivity in plants and animals	●	●	●	●	●	●	●	●	●	●	●	●
11.2	Mammal sense organs			●		●	●				(●)	●	●
11.3	The eye	●	●	●	●	●	●	●		●	(●)	●	●
11.4	The ear			●		●			(●)				
12	**Coordination and response**												
12.1	Information, messages and action	●	●	●	●	●	●	●	●	●	●	●	●
12.2	Mammal nervous system			●	●	●	●	●	●	●	●	●	●
12.3	Nervous impulses			●	●	●			(●)		●		
12.4	Reflex action	●	●	●	●	●	(●)	●	●	●	●	●	●
12.5	Learned behaviour					●					●	●	
12.6	Instinctive behaviour					●					●		
12.7	The brain			●		(●)			●	●	●	●	
12.8	Misused drugs			●		●			(●)				●
12.9	Alcohol—ethanol	●	●	●		●	(●)	●	●	●	(●)		●
12.10	Endocrine system	(●)	(●)	(●)	(●)	(●)	(●)	●	(●)	●		●	(●)

GROUPS AND SYLLABUSES

	LEAG A	LEAG B	LEAG C	MEG A	MEG B	NEA	NISEC	SEG	WJEC	SEB Standard Grade*	SEB Ordinary Grade	MEG Combined Science (Dual Award)
Coordination and response (*cont*)												
12.11 Nervous and hormonal systems compared			(●)	●	●	●	●	●	●			
12.12 Feed-back							●					
12.13 Taxis					●	●				●	●	
12.14 Tropisms	●	●	●		●	●	●	●	●	●	●	●
12.15 Geotropism					●					●	●	
12.16 Photoperiodism	●	●								●		
13 Support and locomotion												
13.1 Principles of support	●	●	●	●	●		●	●			●	●
13.2 Support in plants			●	●	●			●			●	●
13.3 Support and locomotion in animals		●			●			●			●	●
13.4 Principles of movement		●	●		●			●		●	●	●
13.5 Mammal tissues for support and locomotion				●	●	(●)		●	●	●	●	●
13.6 Mammal skeleton					(●)			(●)		(●)		●
13.7 Limbs and limb girdles of Man	(●)	(●)	(●)	(●)	(●)		(●)		(●)	●	●	●
13.8 Joints	●	●	●	●	●	●	●		●	●	●	●
13.9 Movement of an arm	●	●	●	●	●	●	●	●	●	●	●	●
13.10 Functions of mammal skeletons					●		(●)	●	●		●	●
13.11 Sports injuries				(●)						●		
14 Reproduction: mainly plants												
14.1 Asexual and sexual reproduction compared			●	●	●	●	●	●	●	●	●	●
14.2 Asexual methods of reproduction			●	●	●	●	●		●	●	●	●
14.3 Surviving winter—perennation	●		(●)	(●)	(●)	(●)	(●)		(●)	(●)	(●)	(●)
14.4 Vegetative propagation				●	●	●	●	●	●	●	●	(●)
14.5 Flowers	●	●	●	●	●	●	●	●	●	●	●	●
14.6 Self and cross-pollination	●	●	●	●			●	(●)	●	●		
14.7 Wind and insect pollination					●	●	●		●	●		
14.8 Fertilization and its consequences	●	●	●	●	●		●	●	●		●	●
14.9 From flower to seed	●	●	●	●			●	●	●		●	
14.10 Fruits				●	●	●		●	(●)	●		●	(●)
15 Reproduction: humans												
15.1 Sexual reproduction in humans	●	●	●	●	●	●	●	●	●	●	●	●
15.2 Placenta	●	●	●	●	●	●	●	●	●	●	●	●
15.3 Menstrual cycle	●	●	●	●	●	●	(●)	●	●		●	(●)
15.4 Contraception				●	●	●	●					●
15.5 Sexually transmitted diseases (STD)	●	●			●		(●)	(●)	●			●
15.6 Abortion					(●)				●	(●)		●
16 Growth of cells and populations												
16.1 Principles of growth				●	●	●	●	●		●	●	●
16.2 Factors affecting growth				●	●	●	●	●		●	●	
16.3 Human growth				●	●			●				
16.4 Seed structure and germination				●	●	●	●	●			●	●
16.5 Conditions necessary for germination	●		●	●	●	●	●	●	●		●	
16.6 Growth measurement and its difficulties			●							●		
16.7 Growth of populations				●	●	●	●			●		
16.8 Human population				●	●	●	●			●		
16.9 Population structure by age and sex				●								
17 Genes, chromosomes and heredity												
17.1 The nucleus, chromosomes and genes	●	●	●	●	●	●	●	●	●	●	●	●
17.2 Genes and characteristics	●	●	●	●	●	●	●	●	●	●	●	●
17.3 Human blood groups: co-dominance				●	●	●		●	●	●		●
17.4 Mendel's experiments	●	●	●	●	●	●	●	●	●	●	●	●
17.5 Hints on tackling genetic problems	●	●	●	●	●		●	●	●			●
17.6 Back-cross test	●	●	●	●	●		●	●	●			●
17.7 Ratios of phenotypes	●	●	●	●	●		●	●	●			●
17.8 Sex determination in mammals	●	●	●	●	●	●	●	●	●	●	●	●
17.9 Sex linkage							●			(●)		(●)
17.10 Mitosis and meiosis in the life cycle	●			●	●	●	●	●	●	●		●
17.11 How chromosomes move apart at cell division				●	●	●	●	●	●			●
17.12 Mitosis and meiosis compared				●	●	●	●	●	●			●
17.13 Meiosis shuffles genes				●	●	●	(●)	(●)	(●)			●
17.14 Variation in populations	●	●	●	●	●	●	●	●	●	●	●	●
17.15 Mutation	●	●	●	●	●	●	●	●	●	(●)	●	(●)
17.16 Genetic engineering				●		●	(●)		●		●	●
18 Evolution												
18.1 Selection of the 'best' from a variety	●	●	●	●	●	●	●		●	●		●

GROUPS AND SYLLABUSES	LEAG A	LEAG B	LEAG C	MEG A	MEG B	NEA	NISEC	SEG	WJEC	SEB Standard Grade*	SEB Ordinary Grade	MEG Combined Science (Dual Award)
Evolution (cont)												
18.2 Examples of natural selection	●	●	●	●	●	●	●		●	(●)		(●)
18.3 Evolution by natural selection			●	(●)	●	●	●		●	(●)		●
18.4 Charles Darwin (1809–82)				●	●		●		●			
18.5 Evidence for evolution					●							
18.6 Artificial selection	●	●	●	●	●	●	●	●	●	●		●
19 Ecology												
19.1 The biosphere–its limits and organization	●	●	●	●	●	●	●	●	●	●	●	●
19.2 Food chains, food webs and food cycles	●	●	●	●	●	●	●	●	●	●	●	●
19.3 Feeding relationships between species	●	●	●	(●)	●	●	●	●	●	●	●	●
19.4 Stable and unstable ecosystems		(●)	(●)		●			●		(●)	●	●
19.5 Pond ecosystem		●	(●)					(●)		(●)		(●)
19.6 Woodland ecosystem	●		(●)	(●)	(●)			●	(●)	(●)		
19.7 Soil ecosystem	●		●	●	●			●	(●)			
19.8 Keys	●	●	●	●	●	●	●	●	●			
19.9 Soil components	●		●	●	●			●	(●)			
19.10 Nitrogen cycle	●	●	●	●	●	●	●	●	●	(●)	●	●
19.11 Carbon cycle	●	●	●	●	●	●	●	●	●	(●)	●	●
19.12 Earthworms and soil	●						●				●	
19.13 Water cycle	●		●	●	●	●	●	●	●	(●)		
20 Man and his environment												
20.1 Ploughing										●	●	●
20.2 Liming and fertilizing				●		●	●	(●)		●	●	●
20.3 Crop rotation	●	●	●	●	●		●		●	(●)	●	●
20.4 Pest control					●			●		●	●	●
20.5 Human population crisis (problems)	●	●	●	●	●	●	●	●	●			●
20.6 Pollution	(●)	(●)	●	●	●	●	●	●	●	(●)	●	(●)
20.7 Depletion of resources			●	●	●	●	●	●	●	●	●	●
20.8 Human population crisis (solutions)	●	●	●	●	●			●	●	●	●	●
20.9 Types of disease in Man												(●)
20.10 Natural defences of the body against pathogens	●	●	●	●	●	●		(●)	●		●	●
20.11 Notable contributors to health and hygiene								(●)				
20.12 Options for a human future	●	●	●	●	●			●	●	●		●
21 A variety of life												
21.1 Algae	●	●	●		●		●		●	(●)		●
21.2 Mosses and ferns					(●)		●		●			
21.3 Flowering plants				●	●		●		●		●	●
21.4 Annelids	●								●			
21.5 Molluscs	●	●			●				●			
21.6 Crustacea			●	●	●							
21.7 Insects			●		●			●	●	●	●	(●)
21.8 Locust											(●)	
21.9 House-fly		OR			OR						OR	
21.10 Large cabbage white butterfly	●				OR					●		
21.11 Honey bee										(●)		
21.12 Mosquito							●					
21.13 Malaria and other mosquito-borne diseases				●					●			
21.14 Importance of insects to Man					(●)							(●)
21.15 Pork tapeworm					●		●		●		(●)	
21.16 Bony fish		●			●				(●)	(●)	(●)	
21.17 Amphibia		●									●	
21.18 Birds	●								●	(●)	(●)	
21.19 Mammals	●								●			
21.20 Rabbit	OR											
21.21 Bank vole												
22 Biology as a science												
22.1 Scientific method	●	●	●	●	●	●	●	●	●	●	●	●
22.2 Reporting your own experiments	●	●	●	●	●	●	●	●	●	●	●	●
22.3 Scientific units of measurement	●	●	●	●	●	●	●	●	●	●	●	●
22.4 Elements, compounds and mixtures	●	●	●	●	●	●	●	●	●	●	●	●
22.5 Energy	●	●	●	●	●	●	●	●	●	●	●	●
22.6 Surface area to volume ratio	(●)	(●)	●	(●)	●	●	●	●	●	(●)	(●)	●
22.7 Handling measurements and making them meaningful	●	●	●	●	●	●	●	●	●	(●)	(●)	(●)
22.8 Drawings	●	●	●	●	●	●	●	●	●	●	●	●
22.9 Ideas for experiments of your own	●	●	●	●	●	●	●	●	●	●	●	

* Proposed syllabus

Examination Boards: Addresses

NORTHERN EXAMINATION ASSOCIATION (NEA)

JMB Joint Matriculation Board
Devas Street, Manchester M15 6EU

ALSEB Associated Lancashire Schools Examining Board
12 Harter Street, Manchester M1 6HL

NREB North Regional Examinations Board
Wheatfield Road, Westerhope, Newcastle upon Tyne NE5 5JZ

NWREB North-West Regional Examinations Board
Orbit House, Albert Street, Eccles, Manchester M30 0WL

YHREB Yorkshire and Humberside Regional Examinations Board
Harrogate Office — 31–33 Springfield Avenue, Harrogate HG1 2HW
Sheffield Office — Scarsdale House, 136 Derbyshire Lane, Sheffield S8 8SE

MIDLANDS EXAMINING GROUP (MEG)

Cambridge University of Cambridge Local Examinations Syndicate
Syndicate Buildings, 1 Hills Road, Cambridge CB1 2EU

O & C Oxford and Cambridge Schools Examinations Board
10 Trumpington Street, Cambridge CB2 1QB *and* Elsfield Way, Oxford OX2 8EP

SUJB Southern Universities' Joint Board for School Examinations
Cotham Road, Bristol BS6 6DD

WMEB West Midlands Examinations Board
Norfolk House, Smallbrook Queensway, Birmingham B5 4NJ

EMREB East Midlands Regional Examinations Board
Robins Wood House, Robins Wood Road, Aspley, Nottingham NG8 3NR

LONDON AND EAST ANGLIAN GROUP (LEAG)

London University of London Schools Examinations Board
Stewart House, 32 Russell Square, London WC1B 5DN

LREB London Regional Examining Board
Lyon House, 104 Wandsworth High Street, London SW18 4LF

EAEB East Anglian Examinations Board
The Lindens, Lexden Road, Colchester, Essex CO3 3RL

SOUTHERN EXAMINING GROUP (SEG)

AEB The Associated Examining Board
Stag Hill House, Guildford, Surrey, GU12 5XJ

Oxford Oxford Delegacy of Local Examinations
Ewert Place, Summertown, Oxford OX2 7BZ

SREB Southern Regional Examinations Board
Avondale House, 33 Carlton Crescent, Southampton, SO9 4YL

SEREB South-East Regional Examinations Board
Beloe House, 2–10 Mount Ephraim Road, Tunbridge TN1 1EU

SWEB South-Western Examinations Board
23–29 Marsh Street, Bristol, BS1 4BP

WALES

WJEC Welsh Joint Education Committee
245 Western Avenue, Cardiff CF5 2YX

NORTHERN IRELAND

NISEC Northern Ireland Schools Examinations Council
Beechill House, 42 Beechill Road, Belfast BT8 4RS

SCOTLAND

SEB Scottish Examinations Board
Ironmills Road, Dalkeith, Midlothian EH22 1BR

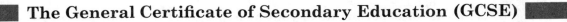

THE GCSE AND SCE EXAMINATIONS

The General Certificate of Secondary Education (GCSE)

The GCSE in the science subjects has certain clear aims.

1 To give credit for what a student knows, can do and can reason out from facts presented to him. In Biology this means there is less emphasis on book learning and more on understanding through doing set practical work and even designing one's own experiments. Teachers assess the experimental work throughout the course, contributing as much as 20% of the final marks. The remaining marks are gained by sitting examination papers.

2 To emphasize the importance of the sciences to everyday life both now and into the future. Many of the topics covered in biology syllabuses affect our everyday lives: health and sex education are examples. Agriculture and medicine, both offshoots of Biology, also influence our lives. Man's effect on his environment will affect future generations.

3 To promote an interest in science. The GCSE is not just intended as a preparation for further study, say at 'A' level and beyond. The courses are intended to be interesting in themselves and part of the purpose of Biology is 'to develop an interest in and enjoyment of, the study of living organisms'. Not only should this study make students more aware of themselves and their environment but they should become used to observing and to making deductions of their own. Statements on scientific matters made in the news media will have to be carefully prepared if they are to pass the critical judgement of GCSE-trained pupils.

4 To rank the performance of 16-year-olds in England, Wales and Northern Ireland on a continuous scale. In the past the GCE attempted to rank the top 20% of pupils and the CSE ranked the next 40% (with some overlap). These examinations were set and marked by a large number of Examination Boards whose examinations were not necessarily of the same standard. Now that there are only five Examining Groups to set exams in each subject, comparability is better. This should make it easier for employers to compare the standard of one applicant with another by looking at their GCSE grades.

The Scottish Certificate of Education (SCE)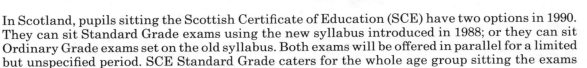

In Scotland, pupils sitting the Scottish Certificate of Education (SCE) have two options in 1990. They can sit Standard Grade exams using the new syllabus introduced in 1988; or they can sit Ordinary Grade exams set on the old syllabus. Both exams will be offered in parallel for a limited but unspecified period. SCE Standard Grade caters for the whole age group sitting the exams — not just the top 30% or so as in Ordinary Grade.

There are three *levels* of 'presentation' (entry) for the exam: Foundation (F), General (G) and Credit (C); however, in Biology only G and C levels will be offered. There are seven *grades* of attainment: Standard 1 is the best and Standard 7 means that a pupil, at the very least, has completed the course.

On the Certificate is recorded not only the Standard reached but also the *profile* of the pupil. The profile shows how well the pupil scored in the various 'elements' (skills) stated in the aims of the syllabus. These separate scores are added to give the total which determines the Standard.

As in GCSE, the SCE Standard Grade in Biology stresses both experimentation and the use of data obtained from the experiments of others. It also expects knowledge of the applications of Biology to everyday life, sport and to environmental issues. Assessment includes 'external' exams and 'internal' (teacher) assessment of practical work. Standard Grade has a better chance of achieving its aims in practical work than the GCSE, in that in Scotland class size in science subjects is limited to 20.

This book meets the needs of students by fitting in with the aims of GCSE and SCE. There are four needs it provides for:

1 It emphasizes the **experimental basis for scientific knowledge**. It explains the way Biologists set up experiments. It sets out the ideas in Physics and Chemistry which Biologists have to use — a useful addition for those who are not studying those subjects. It explains the units of scientific measurement, relating these to biological needs and it explains how collected measurements can be interpreted to make sense.

2 It puts the facts deduced by experiment during the course into a **framework of knowledge**. No isolated conclusions from experiments are of much use unless they fit into, and add to, a background of knowledge (see Fig. C).

3 It sets out detailed **advice on both revision and examination techniques**. The student must know **how** to revise for his written exams and **how** to show his knowledge in the best possible way to gain maximum marks. Whereas achievement of the lower grades in GCSE requires simple answers of a few words, phrases or sentences this is not true for grades A and B. Here questions may be shorter but they require more thought and longer answers. In SCE answers need be no longer than a paragraph.

4 It provides **questions** requiring both short and long answers for students to try. Advice on answering them and the answers themselves help students to achieve their best mark in exams by practice beforehand.

5 It provides **ideas for experiments of your own** linked to the revision material.

The use of this book is not limited to its being a revision guide. Many teachers have used past editions as a class text to provide background to a fully experimental course. It is hoped that this edition will continue to fulfil this need too.

1 LIFE

Living things are called **organisms**. Three large groups of organisms are the **green plants**, e.g. grass; **non-green plants**, e.g. mushrooms; and the **animals** (see units 2.3 and 2.4). All organisms perform *all* the seven 'vital functions' at some time during their existence; and their bodies are made of cells. Some organisms remain, for a time, **dormant** (inactive), e.g. as seeds, spores or cysts. These bodies appear not to perform vital functions but are activated by suitable stimulation to do so, e.g. by germinating.

Eight characteristics of organisms (*Mnemonic:* GERMS NR Cs 'germs in our seas')
Growth
Excretion
Respiration
Movement
Sensitivity
Nutrition
Reproduction
Cells

G **Growth:** cells divide and then get larger again by adding more living material (made from their food) until they repeat the process. (See unit 16.1.)

E **Excretion:** removal of waste products from **metabolism** (all the chemical reactions within the body). (See unit 10.)
NB Do not confuse this with 'egestion' (removal of **indigestible** matter – which has thus never entered cells to be metabolized).

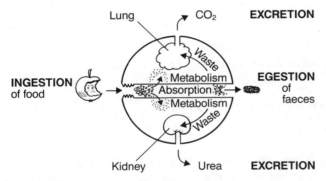

Fig. 1.1 The distinction between excretion and egestion

R **Respiration:** release of energy within cells from food so as to power other vital functions. In most organisms, requires oxygen and releases carbon dioxide and heat. (See unit 9.)

M **Movement:** an animal moves its whole body, using limbs or their equivalent. A plant 'moves' only by *growing* parts of itself towards or away from influences important to it. (See units 12.13 and 12.14.)

S **Sensitivity and response:** influences (**stimuli**) in the surroundings (**environment**) stimulate certain areas of an organism so that they send messages to other parts which respond, e.g. by movement, growth or secretion. (See unit 11.)

N **Nutrition:** intake of food materials from the environment for building up and maintaining living matter. (See unit 4.2.)

R **Reproduction:** formation of more individuals either from one parent (asexually) or two (sexually). (See unit 14.1.)

All organisms eventually die. **Death** is when metabolism ceases completely.

C **Cells:** the simplest units of life. All cells, when young, have at the very least three parts: *a membrane* enclosing jelly-like *cytoplasm*, in which lies a *nucleus* which controls their life. (See unit 1.2.) These three parts make up *protoplasm* (living matter). The cell wall secreted outside the protoplasm, by plant cells only, is non-living. Cells cannot live without supplies of energy, food, water and O_2 and a suitable environmental temperature and pH.

Cells from animals and green plants show differences, as seen in Fig. 1.2.

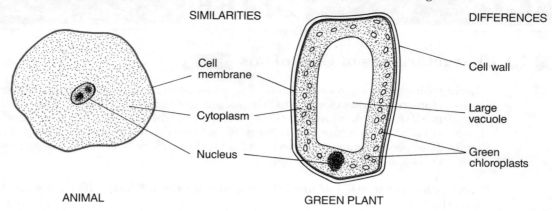

Fig. 1.2 Generalized animal and green plant cells as viewed through a light microscope

1.2 Cells in Detail

1 Observing

Cells need to be stained to show up their parts better under the light microscope.

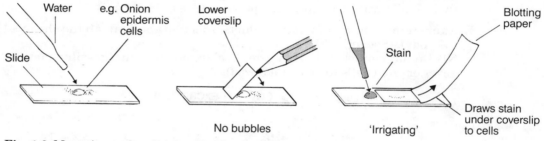

Fig. 1.3 Mounting and staining cells for microscopic examination

Coverslip delays water drying up around cells; permits viewing under high power without distortion under the light microscope.

2 Organelles

Only certain organelles (parts of the cell with special functions) can be seen under the light microscope's magnification. Even smaller organelles can be examined with an electron microscope.

(*a*) **Cell wall**
   Made of cellulose.
   Freely permeable (porous) to all kinds of molecules.
   Supports and protects the cell.
   Supports non-woody plant organs, e.g. leaves, by water pressure within vacuole distending the cell wall.
   Osmoregulates by resisting entry of excess water into cell. (See Fig. 7.6C.)

(*b*) **Cell membrane**
   Exterior of all protoplasm.
   Very thin layer of protein and oil.
   Freely permeable to water and gases only,
   Selectively permeable to other molecules (e.g. allows foods in but keeps unwanted molecules out).

(*c*) **Vacuoles**
   Spaces for various functions, e.g. food storage, osmoregulation.
   Plant cell vacuoles contain 'cell sap' (a weak solution of sugar and salts) inside a membrane.

(*d*) **Cytoplasmic matrix**
   Supports organelles.
   Consistency of raw egg-white.
   Up to 80% water; remainder mainly protein.
   Often contains grains of stored food: starch (plants); glycogen (animals).

(e) **Nucleus**
Stores and passes on cell 'information'.
Contains many long strands of DNA (invisible by light microscope).
When cell divides, DNA coils up to form chromosomes (visible). (See unit 17.1.)
Segments of DNA are called **genes**.
Genes are responsible for characteristics of organisms, e.g. blood group and eye colour. (See unit 1.3.)

(f) **Chloroplasts** (for photosynthesis)
Large bodies containing chlorophyll (green).
Chlorophyll converts sunlight energy into chemical energy (ATP).
ATP is used to combine CO_2 with H_2O making glucose – which stores the energy in its bonds.

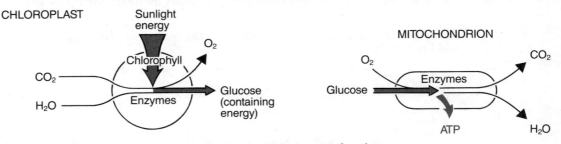

Fig. 1.4 The roles of chloroplasts and mitochondria in transforming energy

(g) **Mitochondria** (for cell respiration)
Just specks ($2\mu m$) under the light microscope.
Absorb O_2 and glucose.
Break down glucose to CO_2 and H_2O. This releases energy from glucose bonds to form ATP.
ATP is chemical energy a cell can use – for *any* vital function (see unit 9.14).

(h) **Ribosomes** (protein factories)
Invisible (20nm) without the electron microscope.
Minute bodies in thousands in cytoplasm.
Assemble amino-acids into proteins, each different according to purpose. (See unit 4.4.)
Instructions for assembly from nucleus.

1.3 How the Nucleus 'Controls' the Cell

Every gene is a recipe for a different protein.
Required recipes are 'copied' and passed to ribosomes.
Ribosomes assemble amino-acids in a special order – according to recipe.
These different proteins are either

1 *secreted*   by gland cells, e.g. digestive enzymes, hormones or
2 *retained*   within cells for metabolism, e.g. enzymes for photosynthesis, respiration; haemoglobin in erythrocytes.

In both cases the proteins determine what each cell can do. So the nucleus, through the proteins it determines, controls what cells can do.

1.4 Sizes of Cells and Their Sub-units

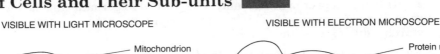

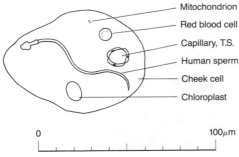

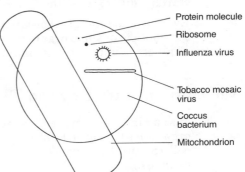

Fig. 1.5 Size

1.5 Enzymes and Metabolism

Enzymes are

1 catalysts – substances that speed up chemical reactions. These reactions do not change the catalyst. So even small amounts of enzyme can do a big job.

2 protein – whose chemical shape (see Fig. 1.6) is special to the substance it works on.

3 specific – starch alone fits into the special shape of the enzyme salivary amylase, not protein or anything else. So starch alone is digested by it.

4 temperature sensitive – boiling destroys enzymes (by altering their shape); cooling only slows down their action. Best (i.e. 'optimum') for mammals is blood temperature.

5 pH sensitive – each enzyme has its own preferred (optimum) pH, e.g. optimum pH for pepsin is pH 2 (acid); for salivary amylase pH 6.8 (almost neutral); for lipase pH 9 (alkaline). (See Fig. 6.5.)

Enzymes catalyse all chemical reactions of the body (metabolism). Without enzymes, reactions would not go fast enough for life to exist.

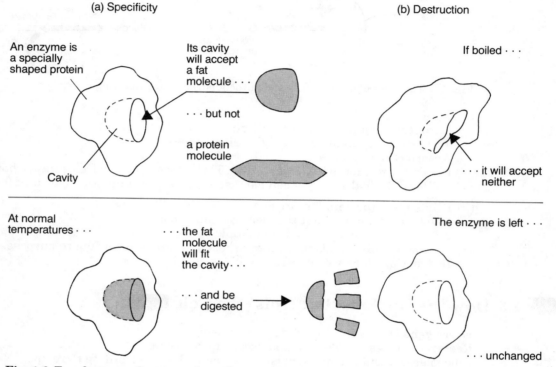

Fig. 1.6 Two features of enzymes dependent on their chemical shape

Metabolism includes

1 anabolism: building up complex molecules, e.g. in photosynthesis, food storage.

2 catabolism: breaking down complex molecules, e.g. in respiration, digestion. It occurs both within cells (e.g. respiration) and outside them (digestion).

Commercially important enzymes

1 'Biological' washing powders – enzymes that digest stains on clothing. They work at warm temperatures – saving expense of boiling, and damage to clothes by boiling or rubbing.

2 'Malting' in beer making – amylases formed in germinating barley digest starch in the grain to malt sugar (maltose). This sugar is then fermented by yeast – whose enzymes cannot use starch – during brewing (see Fig. 3.12).

Every useful biological product – food, flavourings, antibiotics, vaccines etc. – is a product of the metabolism of some organism or another. So, in reality, *every* enzyme involved in making these products is important commercially.

1.6 Units of Life Beyond The Cell

Just as inorganic molecules are built up into organic molecules, which in turn are built into organelles (see unit 1.2), so cells are sub-units of organisms. There is a great variety of types of cell. (See unit 5.5, leaf cells; 7.9, xylem and phloem; 8.2, blood cells; 12.2, neurone; 13.5, bone; 15.1, gametes.)

Tissues are groups of cells, usually of the same type, specialized to carry out certain functions, e.g. muscle for movement, nerve for sending 'messages', xylem for transport and support.

Organs are made up of tissues coordinated to perform certain functions, e.g. eye, leaf, kidney.

Organ systems are groups of organs which combine to perform their functions, e.g. gut, endocrine system, nervous system. The nervous system consists of brain, nerve cord and nerves.

Organisms, depending on their complexity, may each be just one cell, e.g. a bacterium, or *Amoeba*, or millions of cells with a variety of functional units as above, e.g. an oak tree or Man. An organism which reproduces sexually is not much use on its own, unless it self-fertilizes. The basic unit of reproduction is thus usually a **breeding pair**. From this arise **populations** – as small as herds or as large as hundreds of herds occupying an island or a continent. All the populations of this type of organism form a **species** (see Fig. 1.7). Populations of different species living in balance in nature are called **communities**. Communities form part of **ecosystems** in the **biosphere**. (See unit 19.1.)

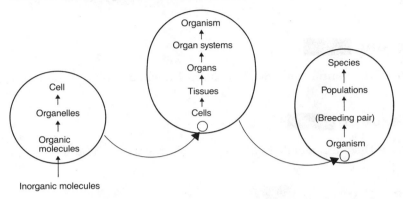

Fig. 1.7 Units of life

2 CLASSIFICATION

2.1 Linnaeus and His Classification System

Carl Linnaeus of Sweden in 1735 introduced the basis of modern **taxonomy** (classification). All species are given two names in Latin – the **binomial system** of naming:

1 genus name, written first, which starts with a *capital* letter, e.g. *Homo* (man).
2 species name, written second, which starts with a *small* letter, e.g. *sapiens* (modern).

The binomial ought to be printed in italics but is underlined when handwritten or typed by scientists, e.g. <u>Panthera</u> <u>tigris</u> (tiger).

Species: a group of organisms capable of breeding to produce fertile offspring. They are very similar, but do show variety.

Genus: a group of organisms with a large number of similarities but whose different sub-groups (species) are usually unable to inter-breed successfully.

2.2 Groups and Sub-groups

Just as species are sub-groups of genera, so Linnaeus grouped genera into larger and larger groups. Each group included as many *similarities* as possible. The largest group is a kingdom, the smallest a species. The lion can be classified as follows:

Kingdom	Animalia	– animals, as opposed to plants.
Phylum	Vertebrata	– animals with backbones (fish, amphibia, reptiles, birds and mammals).
Class	Mammalia	– hairy, warm blooded, suckle young on milk.
Order	Carnivora	– mainly flesh-eating group (cats, dogs, bears, seals).
Family	Felidae	– cats, large and small.
Genus	*Panthera*	– large cats (includes tiger *P. tigris*; leopard, *P. pardus*).
Species	*leo*	– lion only.

Mnemonic: Kadet, P.C., OFficer, General inSpector (promotion in the police force).

Advantages of the system

1 Universal: Japanese, Bantu or Russian biologists all understand that *Felis catus* means 'house cat' without having to resort to a dictionary.

2 Shorthand information: one word, e.g. mammal, conveys a mass of information to all biologists. (See unit 21.19.)

3 Reflects evolutionary relationships: e.g. the five classes of vertebrate are very different (see unit 2.4), yet all have a common body plan. The basic plan (in Fish) (see p. 8), was improved upon, allowing land colonization (Amphibia) and exploitation (Reptiles and Mammals), and even conquering of the air (Birds). The classification of vertebrates thus probably reflects the evolutionary process.

2.3 Plant Kingdom

Classification of the main members of the plant kingdom can be seen on p. 7, Fig. 2.1(a).

2.4 Animal Kingdom

Classification of the main members of the animal kingdom can be seen on p. 8, Fig. 2.1(b).

2.5 Modern Classification

Modern classification systems include four more Kingdoms.

1 Kingdom Protista: this combines the unicellular members of the **Algae** (plants) with the **Protozoa** (animals). This is because some unicells have both plant and animal features, e.g. *Euglena*.
2 Kingdom Bacteria: these unicells have a unique structure with no nucleus. They are minute (mitochondrion size). (See Fig. 3.2.)
3 Kingdom Viruses: not cells, very minute (size of a few genes in a chromosome) and always parasitic inside cells when active. (See unit 3.1.)
4 Kingdom Fungi – instead of Phylum Fungi (see units 3.7 and 3.8). This leaves the Plant Kingdom as meaning entirely *green* plants.

For most purposes Plant and Animal Kingdoms and Viruses (which are not really organisms) are enough.

2.6 Multicell Green Plants and Animals Compared

Table 2.1 Comparison of multicell green plants and animals

	Green plants	*Animals*
Growth	*Branching* – large surface area to absorb nutrients	*Compact* – except for limbs for seeking food
Excretion	*Oxygen* and carbon dioxide Other wastes, e.g. by dropping leaves	Carbon dioxide *Nitrogen* wastes in urine
Reproduction	*Asexual:* frequent	*Asexual:* only simple animals
	Sexual reproduction usual	
Movement	No muscles, cell walls rigid – *Anchored*	Use muscles, skeleton – *Mobile*
Sensitivity and response	No obvious sense organs Response *slow*, using *hormones* to affect growth	*Eyes, ears* etc. obvious Response *fast*, using *nerves* to affect behaviour
Nutrition	*Autotrophic* – inorganic food: synthesized	*Heterotrophic* – organic food: digested
Respiration	*Starch* food store	*Glycogen* food store
	Respiration usually aerobic	
	Anaerobic: produces *ethanol*	Anaerobic: produces *lactic acid*
Cells	*Cell wall, large vacuole, chloroplasts*	All three organelles absent

Note use of mnemonic: 'Germs in our seas' (unit 1.1) as a checklist.

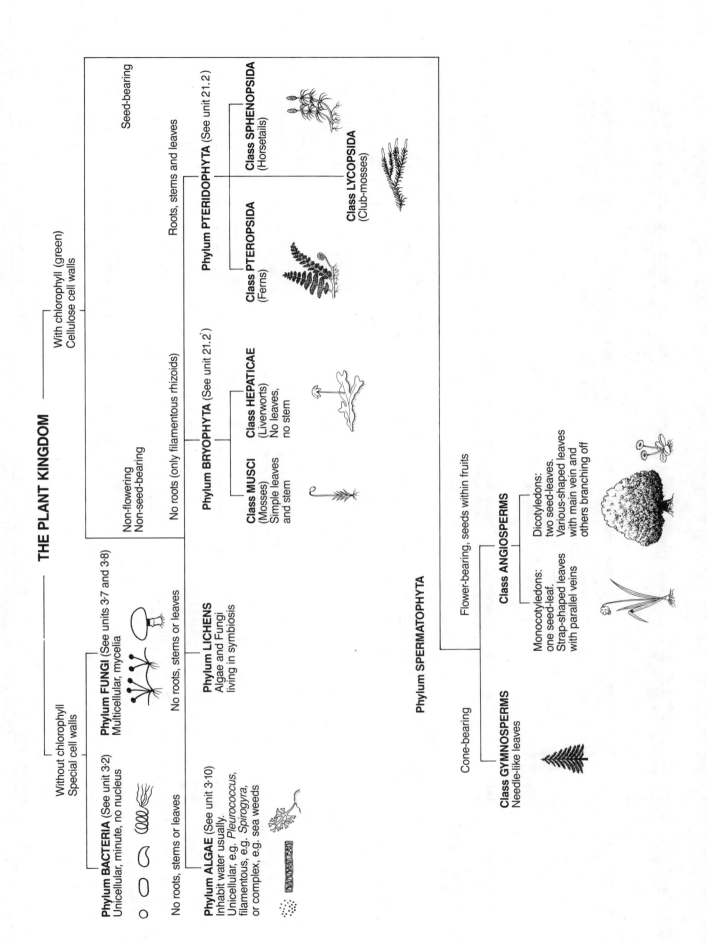

Fig. 2.1(a)

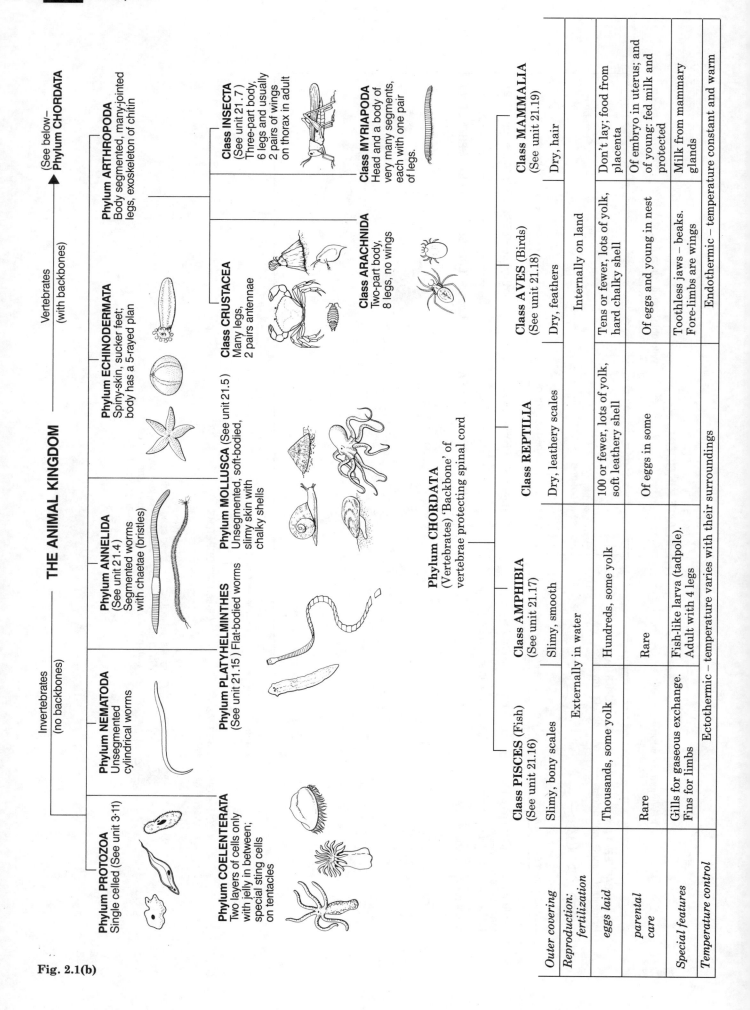

THE ANIMAL KINGDOM

Invertebrates (no backbones)

Vertebrates (with backbones)

(See below – **Phylum CHORDATA**)

Phylum PROTOZOA (See unit 3.11)
Single celled

Phylum COELENTERATA
Two layers of cells only with jelly in between; special sting cells on tentacles

Phylum NEMATODA
Unsegmented cylindrical worms

Phylum PLATYHELMINTHES (See unit 21.15) Flat-bodied worms

Phylum ANNELIDA (See unit 21.4)
Segmented worms with chaetae (bristles)

Phylum ECHINODERMATA
Spiny-skin, sucker feet; body has a 5-rayed plan

Phylum MOLLUSCA (See unit 21.5)
Unsegmented, soft-bodied, slimy skin with chalky shells

Phylum ARTHROPODA
Body segmented, many-jointed legs, exoskeleton of chitin

Class INSECTA (See unit 21.7)
Three-part body, 6 legs and usually 2 pairs of wings on thorax in adult

Class MYRIAPODA
Head and a body of very many segments, each with one pair of legs.

Class CRUSTACEA
Many legs, 2 pairs antennae

Class ARACHNIDA
Two-part body, 8 legs, no wings

Phylum CHORDATA
(Vertebrates) 'Backbone' of vertebrae protecting spinal cord

	Class PISCES (Fish) (See unit 21.16)	**Class AMPHIBIA** (See unit 21.17)	**Class REPTILIA**	**Class AVES** (Birds) (See unit 21.18)	**Class MAMMALIA** (See unit 21.19)
Outer covering	Slimy, bony scales	Slimy, smooth	Dry, leathery scales	Dry, feathers	Dry, hair
Reproduction: fertilization	Externally in water	Externally in water	Internally on land	Internally on land	Internally on land
eggs laid	Thousands, some yolk	Hundreds, some yolk	100 or fewer, lots of yolk, soft leathery shell	Tens or fewer, lots of yolk, hard chalky shell	Don't lay; food from placenta
parental care	Rare	Rare	Of eggs in some	Of eggs and young in nest	Of embryo in uterus; and of young: fed milk and protected
Special features	Gills for gaseous exchange. Fins for limbs	Fish-like larva (tadpole). Adult with 4 legs		Toothless jaws – beaks. Fore-limbs are wings	Milk from mammary glands
Temperature control	Ectothermic – temperature varies with their surroundings	Ectothermic – temperature varies with their surroundings	Ectothermic – temperature varies with their surroundings	Endothermic – temperature constant and warm	Endothermic – temperature constant and warm

Fig. 2.1(b)

3 VIRUSES, MICROORGANISMS, FUNGI AND BIOTECHNOLOGY

3.1 Viruses

Size: between 30-300nm (1/100 size of bacteria) – visible only with electron microscope.

Structure: protein coat around a DNA strand (a few genes). (See Fig. 3.1.)

Living?: no; are not cells, having no metabolism of their own. (See unit 1.1.)

All are *parasites*, killing host cells as they reproduce within them, using the cell's energy and materials. This causes disease, e.g. rabies.

Disease transmission

1 by water, e.g. polio;
2 by droplet (sneezing), e.g. colds, 'flu;
3 by vector (carrier of disease), e.g. mosquito transmits yellow fever and greenfly transmits the TMV (see Fig. 3.1).

Useful: for biological control of rabbits – myxomatosis virus.

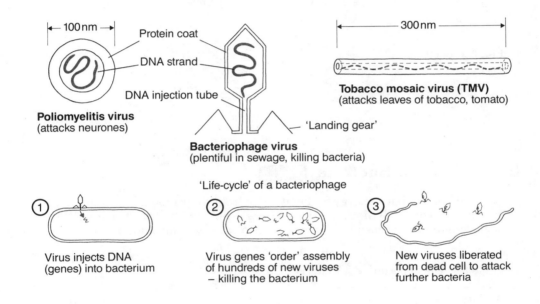

Fig. 3.1 Viruses

3.2 Bacteria

Size: between 0.1-10μm (1/100 size of mammal cheek cell). (See Fig. 1.5.)

Structure: cell is unique in *not* having:
1 nuclear membrane around its single loop chromosome, so there is no nucleus;
2 mitochondria (cell membrane has the same function).

Cell is unlike a green plant cell in having *no*:
1 chloroplasts (therefore bacteria are either saprophytes or parasites) (see unit 4.2);
2 cellulose in cell walls (made of nitrogenous compounds instead) (see Fig. 3.2).

Reproduction

1 asexually: by binary fission, every 20 minutes in suitable conditions.
2 sexually: use a tube to transfer DNA from one bacterium to another.
   Bacteria do *not* reproduce by forming spores. Some bacilli, only, form spores (endospores) for *survival* when conditions become unfavourable.

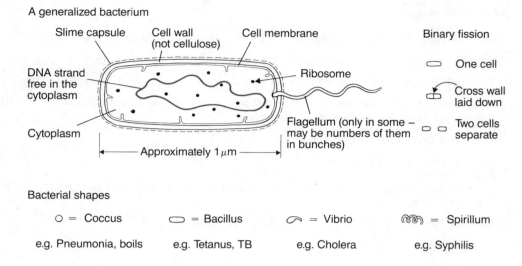

Fig. 3.2 Bacteria

3.3 Importance of Bacteria

Bacteria obtain both energy and materials from their food as we do. In doing so, their metabolism can help or harm Man. In some cases (see nos. 1a, 2a, and 2b below), the *same* bacterium can be both helpful and harmful – according to what it is acting upon.

Table 3.1 The importance of bacteria to Man

Activity	Helpful	Harmful
1 Decomposing	(a) *Dead organisms*, litter, manure into simple nutrients for green plants to use, e.g. CO_2, salts (b) *sewage*, so preventing water pollution	*Food*, e.g. putrefying meat, fish
2 Circulating nitrogen	(a) *Fixing nitrogen*, so increasing soil fertility (see unit 19.10) (b) Converting ammonia (toxic) into nitrate for green plants, i.e. *nitrification*	*Denitrifying* the soil by converting nitrate to nitrogen gas, so reducing soil fertility
3 In industry	(a) *Dairy products:* making butter, cheese, yoghurt (b) *Wineries:* making vinegar (c) Making biogas *fuel* (methane)	*Souring* milk Souring wine *Damaging oil* lubricating engines
4 Affecting health	(a) Producing *antibiotics*, e.g. *Streptomyces* gives over 50 of them (b) Producing human *hormones* by genetic engineering (see unit 17.16)	Causing *disease* in man and his animals (see unit 3.2). Causing *food poisoning*, e.g. *Salmonella*

3.4 How Viruses (V) and Bacteria (B) Reach People

Table 3.2

Method	Examples
1 **By air**	Inhaling minute droplets of infected mucus expelled during sneezing, coughing and even talking. Crowds, poor ventilation and not using handkerchiefs assist infection, e.g. 'flu, colds, measles (V); pneumonia, tuberculosis (B).
2 **By water**	Especially if contaminated by faeces, e.g. poliomyelitis (V); cholera, dysentery (B)
3 **By mud**	Introduced deeply into wounds, e.g. tetanus, gangrene (B)
4 **Sexually** (own species)	During copulation, e.g. genital herpes (V); gonorrhea, syphilis (B) (see unit 15.7). Also AIDS (V) via torn rectums of homosexual men
5 **By vectors** (other species)	Mosquitoes inject yellow fever (V) into blood; houseflies vomit dysentery, cholera (B) onto our food from faeces (see unit 21.9)

In general, viruses and bacteria enter via natural openings of the body which secrete mucus (mucous membranes); or they enter forcibly (nos. 3 and 5), by-passing the natural barrier of the skin.

Infected food and water are particularly common means of entry, e.g. when water is not purified, sewage disposal is poor and personal hygiene is at a low standard (see below).

Hospital hygiene must be particularly good to prevent the great range of available infections from spreading to others in wards and on operating tables (see unit 20.11).

Drug addicts using non-sterile needles add to their misery with infections such as hepatitis, AIDS and blood poisoning.

3.5 Control of Harmful Bacteria

If the basic requirements of bacteria are removed, they die. (See Table 3.3.)

Table 3.3 Control of harmful bacteria

Requirement	Control measure, with examples
1 **Moisture**	**Dried foods:** peas, raisins, milk, meat – keep for ever **Salting:** e.g. ham, or *syruping*, e.g. peaches, plasmolyses (see unit 7.4) bacteria
2 **Organic food**	**Hygiene:** removal of bacterial foods by washing body, clothes, food-utensils; by disposing of refuse, excreta and hospital dressings; cleaning homes
3 **Suitable temperature** (warmth)	**Temperature treatment** (a) *refrigeration:* deep-freeze (− 18°C) suspends life; fridge (+ 4°C) slows rotting to acceptable level; (b) *boiling:* kills most, but not spores; (c) *pressure-cooking* ('sterilizing' or 'autoclaving') for 10 minutes at 10 kN/m² (15 lb/in²) kills all, including spores; (d) *pasteurization* (of milk): heat to 77°C for seconds and rapidly cool to 4°C
4 **Suitable chemical environment**	**Chemicals** are also used to kill bacteria: (a) *chlorine* in drinking water and swimming baths; (b) *disinfectants* in loos; (c) *medical use* of antiseptics, antibiotics, antibodies and drugs in or on Man's body (see unit 20.11); (d) *vinegar* for pickling food (pH too acid for bacteria)
5 **No ultra-violet light**	**Irradiate with ultra-violet light** (thin sliced food, surgical instruments) and plan sunny homes (sunlight contains u.v. light)

3.6 Growing Bacteria

1 Culturing bacteria safely
Golden rules:
(a) Do not culture bacteria except under teacher-supervision.
(b) Do not incubate them unless within taped petri dishes or stoppered test tubes, clearly marked with their source.

(c) Dispose of unwanted cultures by autoclaving.

(d) Take care not to eat during such classes, nor to inhale air close to cultures.

(e) Bacterial 'loops' (of wire), used to transfer bacteria, must be 'flamed' in a Bunsen burner after use.

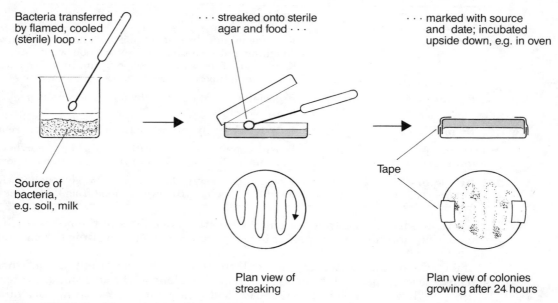

Fig. 3.3 Method for culturing bacteria on agar jelly

2 Testing antibiotics

(a) Particular colonies may be transferred by means of a 'swab' of sterile wet cotton wool on a stick, and 'painted' onto fresh agar to spread it evenly.

(b) Discs of filter paper, soaked in antibiotic, are placed on the culture before incubation, but not on the control.

(c) Clear zones indicate no bacterial growth, i.e. antibiotic effective (see Fig. 3.4).

A doctor may take a swab from an infection to discover, using the method above, which antibiotic is the best for curing it.

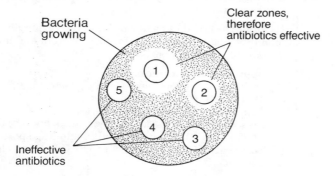

Fig. 3.4 Testing antibiotics

3.7 Fungi – Moulds and Their Culture

Consist of multicellular filaments called hyphae. Cells forming hyphae lack chloroplasts, cellulose and complete cross-walls – in contrast to green plants.

Nutrition saprophytic or parasitic (see unit 19.3).

Distribute themselves by spores, formed asexually.

Moulds

Rhizopus (mould on bread) and *Mucor* (mould on dung) are both 'pin-moulds' (see Fig. 3.5).

Structure: the cytoplasm, with many nuclei in it, lines the cell wall – a continuous tube (of chitin) with no partitions forming separate cells. Inside the cytoplasm is a continuous vacuole. Threads of fungus (hyphae) make up a mycelium.

Nutrition: saprophytic (see unit 4.2). Rootlet hyphae branch through the food, secreting digestive enzymes and absorbing the soluble products.

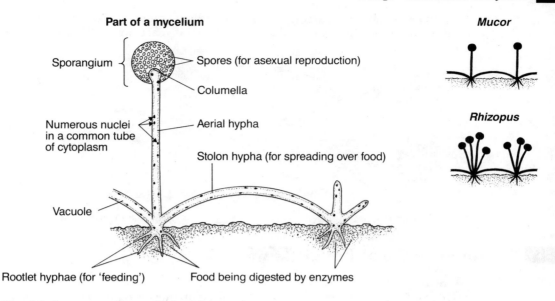

Fig. 3.5 Structure of a mould fungus

Reproduction:

1 **asexually** by hundreds of spores from each sporangium. In *Mucor,* the sporangium wall dissolves in moisture and spores are distributed in a slime-drop by rain or animals. In *Rhizopus*, the wall cracks open when dry and wind distributes dry spores.

2 **sexually** using special hyphae from two different strains to bring together gamete-nuclei. These gametes fuse and a tough zygospore is formed to survive unfavourable conditions, e.g. drought or the winter.

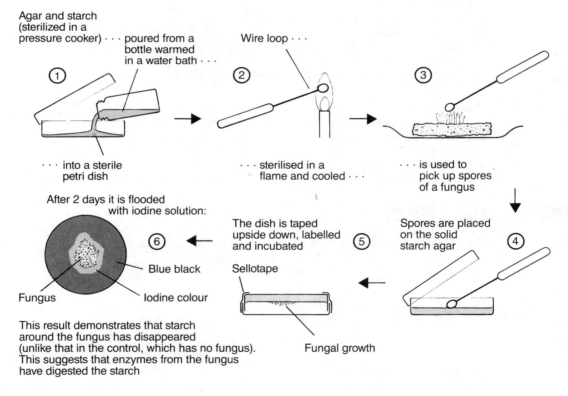

Fig. 3.6 Culturing a mould and testing for saprophytic nutrition

3.8 Fungi – Mushrooms and Yeasts

Mushrooms and toadstools exist unseen as a mycelium within soil, dead wood etc. In damp cool conditions (e.g. October), hundreds of hyphae grow solidly up together, out of the soil, to form a toadstool (see Fig. 3.7). On the underside this sheds millions of spores from 'gills' or pores. Such fungi are usually saprophytic; others are parasites and symbionts (see unit 19.3).

Yeasts are exceptional among the Fungi in being single celled – no hyphae. Natural yeasts on fruit skins ferment them to produce wine. Special yeasts are cultivated by man for brewing, SCP etc. (see Fig. 3.8).

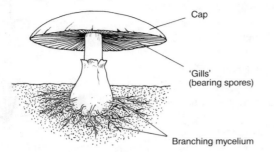

Fig. 3.7 Mushroom fungus

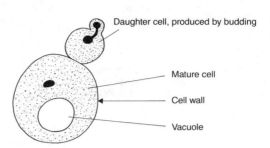

Fig. 3.8 Yeast cells

3.9 Importance of Fungi

Helpful
1 **Decay fungi** (decomposers) release nutrients for green plants from dead organisms.
2 **Yeasts**, respiring anaerobically, provide:

(a) alcohol for brewers and wine-makers: this may be distilled to make spirits, e.g. gin;
(b) CO_2 for bakers (yeast acts on sugar in dough, making it rise)
$$C_6H_{12}O_6 \longrightarrow 2C_2H_5OH + 2CO_2$$
(c) the yeast cells themselves also yield extracts (e.g. 'Marmite'), rich in vitamin B;
(d) 'gasohol' is ethanol brewed from sugar cane in Brazil and distilled to fuel over a million cars there.

3 **Antibiotic-producers**, e.g. *Penicillium* produces pencillin.

4 Food
(a) natural, e.g. mushrooms, chanterelle, truffles.
(b) certain yeasts, grown in fermenter vessels containing solutions of ammonium and other salts which is bubbled through with natural gas, are harvested to give 'single cell protein' (**SCP**), very cheaply. Fed to cattle for quick growth.
(c) the mould *Fusarium graminearum* promises to provide a high quality **mycoprotein** food for humans.

Harmful
1 **Decay fungi** spoil food, e.g. *Rhizopus, Penicillium* on bread, cakes and jam.
2 **Plant diseases**, e.g. potato blight, caused millions to die in the Irish potato famine; 'rust' fungi damage cereal crops seriously; dutch elm disease killed millions of elm trees.
3 **Dry rot** fungus destroys house timbers.
4 **Minor animal diseases**, e.g. 'athlete's foot' and 'ringworm' (both attack the skin).

3.10 Algae

The Algae are green plants with no roots, stems or leaves.
Their cells are little specialized – apart from gametes.
Nutrition is holophytic (see unit 4.2).

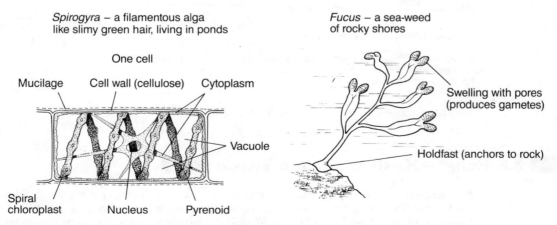

Fig. 3.9 Examples of multicellular algae. See also *Cladophora* (unit 21.1)

The unicellular algae, e.g. *Pleurococcus*, diatoms and desmids (see unit 21.1), are often included with the Protozoa in the Protista.

Importance of algae

1 Diatoms (unicellular algae) are the main plant component of plankton (phytoplankton). They

(*a*) provide the majority of the world's O_2;
(*b*) are at the base of most marine food chains.

2 Some **sea-weeds** are eaten, e.g. 'Irish moss'.
3 Extracts: 'agar' for bacterial culture methods; 'alginates' for ice-cream.

3.11 Protozoa

The Protozoa are single celled animals. Organelles for movement include pseudopodia (for flowing along), flagella and cilia (for swimming).
  Nutrition is holozoic or parasitic.

Amoeba is a large fresh-water protozoan (up to 1 mm in diameter) (see Fig. 3.10).

Locomotion: cytoplasm in the centre (plasmasol) flows forward forming a pseudopodium. At the front the plasmasol fountains out, solidifying to a jelly-like tube (plasmagel) through which the centre flows. The plasmagel re-liquefies at the rear end, flowing into the centre.
Nutrition: holozoic (see unit 4.2). Pursues prey (algae, bacteria, other protozoa) by following the trail of chemicals they exude (chemotaxis, see unit 12.13). The prey is ingested using pseudopodia; digested in a food vacuole; indigestible matter, e.g. cellulose, is egested.
Osmoregulation: water entering continually by osmosis is channelled to the contractile vacuole. When full, this bursts, squirting water out. The process uses energy.
Respiration: gaseous exchange (O_2 in, CO_2 out) occurs over the whole surface area (see unit 9.7).

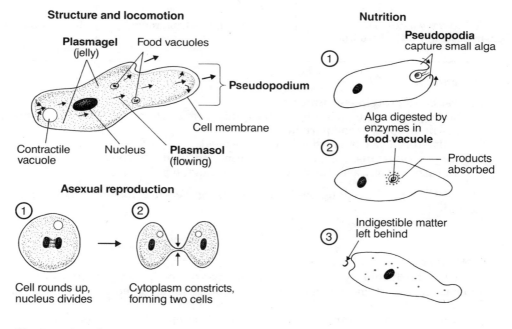

Fig. 3.10 Amoeba

Reproduction: asexually by binary fission; no sexual method.
Sensitivity: moves towards food; away from strong light, harmful chemicals and sharp objects.

Importance of protozoa

1 Malaria parasite (*Plasmodium*), transmitted by mosquito, kills millions of people by fever unless protected by drugs, like quinacrine.
2 Sleeping sickness parasite (*Trypanosoma*), transmitted by tsetse fly, kills millions of people, cattle and pigs in Africa. No drug protection against some types.
3 Dysentery parasite (*Entamoeba*), transmitted by house-fly, causes dysentery (intestinal bleeding and upsets) and liver abscesses.

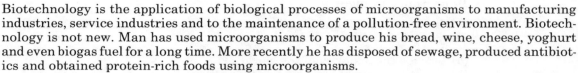

3:12 Biotechnology

Biotechnology is the application of biological processes of microorganisms to manufacturing industries, service industries and to the maintenance of a pollution-free environment. Biotechnology is not new. Man has used microorganisms to produce his bread, wine, cheese, yoghurt and even biogas fuel for a long time. More recently he has disposed of sewage, produced antibiotics and obtained protein-rich foods using microorganisms.

The most recent advances result from understanding cell processes (dependent on enzymes); and from putting useful genes into microorganisms (genetic engineering) to do special jobs.

Biotechnology could transform our lives by providing efficient, low-cost solutions to many of the future problems of an overcrowded world, including shortage of energy and food.

Biotechnology requires cooperation between scientists:

1 Biologists – to find and grow microorganisms and to test the properties of potentially useful ones.
2 Biochemists – to study the metabolism of microorganisms and to genetically engineer them (see unit 17.16).
3 Engineers – to design fermenters that provide the right growth conditions, flows of nutrients and products, and separation of the two at the end.
4 Electronics scientists – to provide sensitive feed-back systems which automatically maintain the chosen conditions for growth.

Examples of the newer aspects of biotechnology are given below:

FOOD – single cell protein (SCP)

Table 3.2 SCP and its sources

Microbial foods	Examples	Notes
Glucose, salts, ammonia	RHM Mycoprotein – from fungi	Human food
Industrial wastes, e.g. molasses, whey, waste wood	Finnish Pekilo – from fungi	10 000 tonnes/year for animals – turns waste into profit, reduces pollution

Advantages of SCP

1 First class protein (see unit 4.6).
2 Can be produced in vast quantities quickly (mass-doubling time can be less than 1 hour).
3 Uses little space – compare farms.
4 Easily stored as dry powder – compare meat.
5 Free of biohazards, e.g. pesticides, hormones, food additives – if handled properly.
6 Can turn costly waste into a valuable product, e.g. Pekilo.

Note: SCP *algae* have long been cultivated, e.g. *Senedesmus* and *Chlorella* in Japan; *Spirulina* in Africa.

ENERGY

Man has traditionally burned recent biomass (wood, dung), which is renewable, and fossil biomass (coal, oil, natural gas), which is not. He is using both far too fast for this to continue long.

Wind, water and solar power are unlikely to supply the whole of world energy requirements.

Nuclear energy is expensive, can provide serious pollution hazards through release of radioactive materials (see unit 20.6) and can only produce electrical energy. This currently cannot be easily stored to fuel automobiles, ships and aeroplanes.

Microorganisms may in the future play a larger role in producing the fuels ethanol, methane and methanol – to substitute for petrol, diesel and liquefied natural gas in vehicles. The chosen microorganisms cannot avoid depending on photosynthesis – which fixes ten times the world's present energy requirements. Ideally the microbes' food should be wastes, e.g. wood, bark, paper. But cultivation of 'fuel plants', e.g. sugar cane and maize (which fix twice as much solar energy as 'ordinary' green plants) for microorganisms to use as food is attractive, especially in the developing countries.

'Gasohol' (ethanol brewed from sugar) fuels over a million vehicles in Brazil; methanol (at present produced chemically) fuels over 5000 vehicles in oil-rich California. Neither fuel provides the pollution hazards of petrol (see unit 20.6). Methane cannot be liquefied cheaply enough for vehicle use, but it can be used direct from 'digesters' in sewage farms to power their pumps and lighting. Small digesters provide many Indian villagers with methane for heating.

SEWAGE DISPOSAL

1 Sewage can contain **pathogens** from infected people, e.g. those suffering from cholera or typhoid. Sewage treatment removes most of these; the rest are killed by chlorination of the water supply.
2 Faeces and urine could cause **eutrophication** (see unit 20.6) if discharged into rivers. Instead, they are broken down in decomposer units by bacteria and fungi. These units are either coke filter beds, where sewage trickles over them, or large sewage tanks bubbled through with air. The mineral salts and carbon dioxide formed by decomposition are absorbed by algae. The decomposers and algae feed a variety of protozoa, worms etc – a special artificial ecosystem.
   The flow of treated sewage contains
(*a*) dead organisms ('sludge') – a valuable fertilizer sprayed onto pastures as a black liquid;
(*b*) 'clean' water – discharged into rivers.

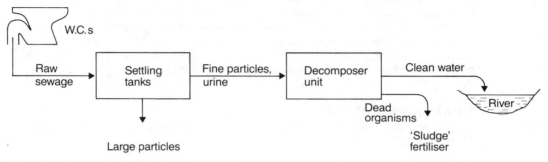

Fig. 3.11 Sewage farm (simplified)

ENZYMES

Enzymes (see unit 1.5) have great advantages over inorganic catalysts.
1 Enormous variety of them, for every purpose.
2 Continuously produced (no shortages or great cost: compare platinum).
3 Work at low temperatures (not costly in energy).
4 Do a precise job, e.g. digesting the lignin of wood, leaving its cellulose intact for useful SCP microorganisms to use – as in the Pekilo process, p. 16.
   Enzymes *secreted* by microorganisms are easier to collect than those kept within their cells. Once extracted they store easily.
   Enzymes are used either in solution, or trapped in polymer beads (where they are less easily destroyed) and can be used over and over again after each harvest of product. Many traditional processes can now be completed more quickly and with better control over the quality of the product using enzymes from microorganisms.

Table 3.3 The use of enzymes

Process	Enzymes from microorganisms used	Purpose
Washing clothes	Proteases	'Biological' washing powders remove stains, e.g. blood
Washing dishes	Amylases	Dish-washer powders remove starch smears on plates
Cheese making	Rennin Lipases	Curdles milk Speed up ripening of Danish blue cheese
Leather making	Proteases	Make leather supple (replace use of enzymes in dog dung!) and remove hair from hides
Brewing	Carbohydrases Proteases	Split starch into maltose and proteins into amino-acids in malt (see Fig. 3.12)

It is likely that in the future chemicals may be produced by a series of fermenters, each containing a single enzyme. By this means the final product can be produced, step by step, and be easily controlled.

Fig. 3.12 Brewing beer (simplified)

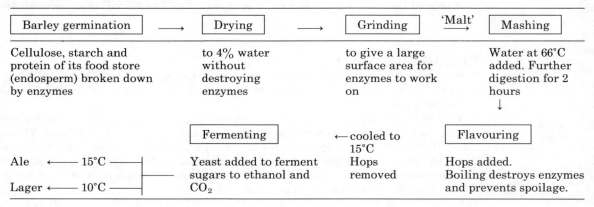

Barley germination	→	Drying	→	Grinding	'Malt' →	Mashing
Cellulose, starch and protein of its food store (endosperm) broken down by enzymes		to 4% water without destroying enzymes		to give a large surface area for enzymes to work on		Water at 66°C added. Further digestion for 2 hours ↓

		Fermenting	←cooled to 15°C	Flavouring
Ale ← 15°C ⎤ Lager ← 10°C ⎦		Yeast added to ferment sugars to ethanol and CO_2	Hops removed	Hops added. Boiling destroys enzymes and prevents spoilage.

Beer brewing in the UK is a **batch process**, i.e. the fermenter is totally cleared of microorganisms and products before a new supply of microorganisms and raw materials are introduced. 'Continuous processing' in beer brewing has been abandoned as it gives unwanted flavours.

A **continuous process** is one where raw materials flow into the fermenter and products and surplus microorganisms are continually harvested, bit by bit. The SCP 'Mycoprotein' is produced in this way. The fermenter is never emptied totally.

HEALTH PRODUCTS

1 Antibiotics are extracted by solvents from certain bacteria and fungi, e.g. penicillin from *Penicillium*.

2 Hormones, e.g. insulin. The human gene for making insulin is inserted by genetic engineering (see unit 17.16) into a bacterium. The engineered bacterium grows rapidly in a fermenter, producing insulin (see unit 12.10).
Advantages:
(a) less costly than extracting insulin from tonnes of slaughterhouse pancreases;
(b) avoids allergic reactions to the animal product.

3 Antibodies. A single lymphocyte cell (see Table 8.1) which is producing a wanted antibody is selected. It is made to fuse with a cancer cell (which can be grown easily outside the body in a fermenter). The single fused cell multiplies rapidly, secreting the antibodies. These can be separated out to give a vaccine.
Advantages:
(a) much more productive than the original method: collecting antibodies from the blood of animals injected with human pathogens;
(b) avoids allergic reactions.

4 Vitamins: bacteria and fungi are usually rich in vitamins for humans – they can be individually extracted. A crude extract from yeast, rich in B vitamins, is 'Marmite'.

PESTICIDES

A number of microorganisms causing diseases of *particular* insect pests are in commercial production. These insecticides do not kill *all* insects as the now banned chemicals DDT and dieldrin did.

MINING

As minerals run out it will become important to recover metals even from low-grade ores and 'tailings' of previous mining operations. Already use is made of bacteria that change copper, zinc, lead, cobalt and uranium into soluble salts. These salts can be washed out from ores containing as little as 0.01% of the metal. This makes extraction profitable.

4 FOODS AND FEEDING

4.1 Food

Food (material for building up protoplasm) is of two types:
1 Inorganic: (simple molecules common to non-living matter), e.g. carbon dioxide, mineral salts and water.
2 Organic: (complex, carbon-containing compounds), e.g. carbohydrates, fats, proteins and vitamins. These classes of molecules are characteristic of living matter.

4.2 Holophytic, Holozoic and Saprophytic Nutrition Compared

There are two fundamentally different methods of nutrition:
1 Autotrophic organisms (plants containing green chlorophyll) need *only inorganic food* from which they synthesize organic molecules, using *energy trapped from sunlight* to drive the reactions.
2 Heterotrophic organisms (animals, fungi, bacteria) have to feed on ready made *organic food*. From this they derive their *energy, released by respiration*. They also need some inorganic food.

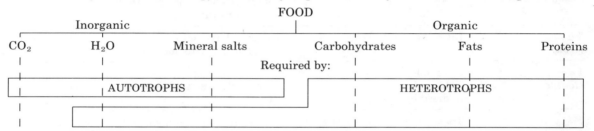

Organic food can be obtained from living organisms (**holozoic** nutrition) or from dead matter (**saprophytic** nutrition). (For other variations see unit 6.1.)

Table 4.1 Comparison of types of nutrition

Type of nutrition	Autotrophic	Heterotrophic	
	Holophytic	*Holozoic*	*Saprophytic*
Examples of organisms	Typical green plants, e.g. *Spirogyra* (unit 3.10) and angiosperms	Typical animals, e.g. *Amoeba* (unit 3.11) and mammals	Bacteria and fungi of decay, e.g. *Mucor* (unit 3.7) and mushrooms
Type of food	Inorganic only: CO_2, H_2O and mineral salts	Organic, H_2O and mineral salts	Dead organic, H_2O and mineral salts
How the food is used	1 CO_2 and water are combined in **photosynthesis** to make carbohydrates 2 carbohydrates are modified and also often combined with salts to **form other organic molecules**, e.g. protein	Food organisms are killed; *ingested* into a **gut**; *digested* by enzymes secreted internally; soluble products *absorbed*; indigestible waste *egested* (eliminated)	Dead organisms or excreta are digested by enzymes secreted **externally** onto them; soluble products absorbed
Source of energy for vital functions	**Sunlight** – trapped by chlorophyll during photosynthesis	Cannot trap sunlight energy since they lack chlorophyll. Rely on **respiration** of organic molecules (the bonds of which contain energy)	

Thus the kinds of organism practising these three forms of nutrition provide food for each other:

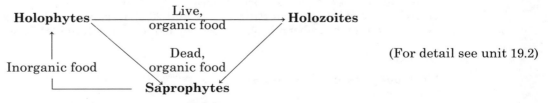

(For detail see unit 19.2)

The principles outlined in Table 4.1 are best studied in detail by reference to simple organisms such as those named (see units 3.7–3.11). Complex organisms, however, have complex requirements and uses for the molecules they absorb. This is taken into account below.

INORGANIC FOOD

1 **Water** (see unit 7.1).
2 **Mineral salts**

4.3 Mineral Salts for Mammals and Angiosperms

(Angiosperms are flowering plants)

Table 4.2 Mammal requirements (especially Man)

Element	Good sources	Uses	Deficiency effects
Ca *(Calcium)*	Cheese; milk; bread (chalk added by law)	Bones and teeth are about $\frac{2}{3}$ calcium phosphate	Brittle bones and teeth
P *(Phosphorus)*	Milk	Bones and teeth; ATP the energy molecule (unit 9.14); DNA – genes and their functions (unit 1.2)	As above
Fe *(Iron)*	Liver, egg yolk	Part of haemoglobin, the oxygen-carrying molecule	Anaemia (lack of red blood cells)
I *(Iodine)*	Sea foods Table salt (iodized by law)	Part of thyroxine, the hormone controlling metabolic rate (unit 12.10)	Goitre – thyroid swelling in adults
F *(Fluorine)*	Toothpaste or tap water that have been fluoridated	Ensures hard tooth enamel, therefore less tooth decay (caries)	Dental caries more likely
Na *(Sodium)*	Table salt (NaCl)	A correct balance of these is required, particularly for proper function of nerves and muscles	
K *(Potassium)*	Plant food		

Table 4.3 Angiosperm requirements (See unit 5.8)

Element	Sources	Uses	Deficiency effects
N *(Nitrogen)*	Nitrates	Protein and DNA synthesis	Poor growth – little protoplasm made
S *(Sulphur)*	Sulphates		
Ca *(Calcium)*	Lime ($CaCO_3$)	'Gum' (middle lamella) between adjacent cell walls	Faulty cell division
Fe *(Iron)*	Iron salts	Enzymes for making chlorophyll	Pale leaves (chlorosis)
Mg *(Magnesium)*	Magnesium salts	Part of chlorophyll molecule	
P *(Phosphorus)*	Phosphates	ATP (energy molecule) in photosynthesis and respiration; DNA synthesis	Poor growth – little energy for synthesis of protoplasm
K *(Potassium)*	Potassium salts	Functions not clear	Poor growth – dehydration

Trace elements include zinc (**Zn**), copper (**Cu**) and manganese (**Mn**). Required in very minute quantities for healthy growth (larger quantities are often poisonous).

Although green plants absorb mineral salts as ions, Man does not always give them to his crops by way of inorganic fertilizers (see unit 20.2). Organic fertilizers, such as dung, also yield salts, once they have been broken down by bacteria and fungi (see units 3.3, 3.9, 19.10).

ORGANIC FOOD

1 **Carbohydrates, fats and proteins**
2 **Vitamins**

4.4 Carbohydrates, fats and proteins

Table 4.4 Carbohydrates, fats and proteins

	Carbohydrates	*Fats* (solids), *oils* (liquid)	*Proteins*
Elements	C, H, O Ratio of H:O is 2:1 (as in H_2O)	C, H, O Ratio of H:O is very high, i.e. very little O	C, H, O, N, often S
Examples	Glucose, $C_6H_{12}O_6$ Starch $(C_6H_{10}O_5)n$	Mutton fat: $(C_{57}H_{110}O_6)$	Haemoglobin, amylase, insulin $C_{254}H_{377}N_{65}O_{75}S_6$
Units	Mono-saccharides (simple sugars, like glucose)	Glycerol + fatty acids	Amino-acids
	These are the smallest units into which these three classes of food can be broken down by digestion (*hydrolysis*). The units can be reassembled into larger molecules again by *condensation*, e.g. when food needs to be stored (see unit 6.2)		
Larger molecules	Di-saccharides (2 units), e.g. sucrose, maltose		Di-peptides (*two* linked amino-acids)
	Poly-saccharides (thousands of units), e.g. starch, glycogen, cellulose		Poly-peptides (*many*)
Chemical tests	1 Blue **Benedict's** solution* + **reducing sugar** $\xrightarrow{\text{boiled}}$ *orange* precipitate 2 **Clinistix** + **glucose** $\xrightarrow{\text{cold}}$ mauve or *purple* 3 Brown **iodine** solution + **starch** $\xrightarrow{\text{must be cold}}$ *blue-black*	1 The clear filtrate obtained from mixing **absolute ethanol** with crushed food, when added to an equal quantity of water, gives a *white emulsion*. 2 **Translucency:** when warmed on paper, makes paper permanently translucent ('grease spot')	1 Colourless 40% **NaOH** + protein extract, add 2 drops blue $CuSO_4 \rightarrow$ *mauve* **Biuret** colour **(Biuret test)** 2 **Albustix** and *some* proteins $\xrightarrow{\text{cold}}$ green or (usually) *blue-green*
Functions	**Energy supply** when respired: 17 kJ/g. Used first. Stored as *starch* (green plants) and *glycogen* (animals, fungi). Transported as sugars **Structural:** cellulose cell walls **Origin of other organic molecules:** e.g. sugar + nitrate → amino-acid	**Energy supply** when respired: 39 kJ/g. Used after carbohydrates. Important in flying, migrating and hibernating animals. (More energy per unit mass than glycogen) **Heat insulation:** subcutaneous fat in mammals **Waterproofing:** of skin, fur, feathers **Buoyancy:** e.g. fish larvae in the sea	**Energy supply** when respired: 18 kJ/g. Important in carnivores, otherwise only respired extensively in starvation **Movement:** *muscles* contract; *tendons* connect muscles to bones; *ligaments* connect bone to bone at joints – all are protein **Catalysts:** *enzymes* make reactions of metabolism possible (see unit 1.5) **Hormones** regulate metabolism (see unit 12.10). Many, e.g. insulin, are protein

* *Note:* **Fehling's** solutions I and II, if mixed to give a *royal blue* solution, give similar results to Benedict's solution.

4.5 Vitamins

Vitamins: organic substances (of a variety of kinds) required in *minute* amounts to maintain health of heterotrophs. Autotrophs make all they need.

Lack of a vitamin in the diet results in a *deficiency disease*, e.g. scurvy. A vitamin for one organism is not necessarily a vitamin for another, e.g. Man suffers scurvy from lack of vitamin C but rats do not because they synthesize their own.

Vitamins A and D are *fat soluble* ingested in fats and oils.
Vitamins B and C are *water soluble* and present in other materials.

How much vitamin C in a food?
Find out the volume of liquidized food needed to decolourize a volume of DCPIP solution of known strength. Do the titration again with a standard vitamin C solution instead of food. By comparing the two volumes, the vitamin content of the liquid can be calculated.

Table 4.5 Vitamins

Vitamins	Good sources	Functions	Deficiency diseases
A	Vegetables, butter, egg yolk. Liver oils, e.g. cod-liver oil, contain both A and D	1 Healthy epithelia 2 Part of 'visual purple' in rod cells of retina (unit 11.3)	Susceptibility to *invasion by disease organisms* *Poor night-vision*
D *'sunshine vitamin'*	Butter, egg yolk. (Can be synthesized in the skin from oils irradiated by ultra-violet light)	Regulation of calcium and phosphate absorption from gut and their deposition in bone	*Rickets:* poor bone formation, weak and often deformed, e.g. 'bow legs' in children
B₁ *(thiamine)*	Wholemeal bread		*Beri beri:* weak muscles, paralysis
B₂ *(riboflavin)*	Liver Yeast Marmite	Different roles in metabolism, especially respiration	Sore mouth cavity, tongue
B₃ *(nicotinic acid)*	Wholemeal bread		*Pellagra:* blistered skin diarrhoea
B₁₂ *(cobalamine)*	Liver; extracted now from a mould, *Streptomyces*	Aids formation of red blood cells	*Pernicious anaemia:* lack of red blood cells
C	Citrus fruits, blackcurrants; fresh vegetables and milk	Healing of wounds; strong skin and capillaries	*Scurvy:* capillary bleeding; poor healing of wounds

Test for vitamin C: blue **DCPIP** solution is turned colourless by **vitamin C** solution (and other reducing agents – which may be present in foods as preservatives). DCPIP may turn red if acid foods are added – but bleaching still occurs with vitamin C.

4.6 Diet

A **balanced diet** is one that maintains health. It must provide enough of the following:

1 **Energy** from carbohydrates and fats when respired.
2 **Materials for growth and repair:**
   from proteins to make muscles, enzymes;
   from mineral salts to make bones, red blood cells.
3 **Vitamins** to help run metabolism.
4 **Water** to transport materials; provide a medium in which they react (see unit 7.1)
5 **Fibre** (roughage) to help peristalsis.

Starvation refers to massive lack of food of all kinds.
Malnutrition refers to lack or excess of particular parts of the diet, e.g. obesity (fatness) results from excessive intake of energy-foods – linked to heart disease.
Anorexia (wasting away) results from not eating enough energy foods. See also mineral salt and vitamin deficiency diseases (units 4.3, 4.5).
Balanced diets differ according to *age* (growing or not), *occupation* (energy requirements and water intake to replace sweat differ), *climate* (less energy needed to keep body warm in tropics than in the Arctic) and *sex*.

Quantity and quality of food is important for health:

1 *Protein*: made of 20 different amino-acids, linked into chains. Of these 20, Man cannot make 8 and so must get them from food. Animal protein and SCP (see p.16) are rich in all of them: 'first class protein'.
Plant protein is usually poor, deficient in some amino-acids: 'second class protein'.
Kwashiorkor (wasting of limbs, puffiness of tissues and pot-belly full of fluid) results from lack of first class protein, e.g. in maize-eating Africans.
2 *Fats*: two kinds – saturated (plentiful in animal fats) and unsaturated (plentiful in plant oils). High intake of animal fats, e.g. butter, seems linked with *heart disease* (p.48). Margarines made from plant oils seem safer.
3 *Additives*: The food you eat may not be what it seems. Substances used in its production, e.g. insecticides, and even in its packaging, e.g. metal, may contaminate it. Other substances also, deliberately added by the food industry to make it appear, smell and taste good, in some cases cause concern.
Although most additives seem harmless and give us distinct advantages, some affect a minority

for the worse. In particular some children become **hyperactive** (i.e. very troublesome, with learning difficulties, despite good intelligence) when they eat certain additives.
Another minority of people develop **allergies** to certain additives, e.g. monosodium glutamate. Since January 1986, the European Community have agreed that all substances added to packaged food, including water, must be listed on the packet in descending order of weight. All such additives (other than flavourings) are given an 'E number', e.g. E140 for chlorophyll (used to colour food green).

5 GREEN PLANT NUTRITION

HOLOPHYTIC NUTRITION

Unique features: uses only inorganic food molecules to photosynthesize sugars and synthesize amino-acids.

5.1 Photosynthesis

Photosynthesis makes sugars and the by-product oxygen from CO_2 and water, using the energy of sunlight, trapped by chlorophyll. Occurs in chloroplasts (see unit 1.2).

The simplest equation for photosynthesis is:

$$6CO_2 + 6H_2O \xrightarrow[\text{chlorophyll}]{\text{sunlight energy}} C_6H_{12}O_6 + 6O_2$$

carbon   water   glucose   oxygen
dioxide   (energy-
                                                  rich)

In reality photosynthesis has *two* distinct stages:

1 photolysis – water is split to give

(a) oxygen gas (by-product)
(b) hydrogen for reducing CO_2 $\Big]$ only in the *light*

2 reduction – of CO_2 by hydrogen to form sugars, e.g. glucose. This needs energy (ATP).
Evidence: if heavy isotope of oxygen, ^{18}O, is used to 'label' water fed to plants, all the O_2 given off is ^{18}O and none is normal oxygen, ^{16}O. If the CO_2, and not the water, fed to the plants is labelled with ^{18}O, *none* of the O_2 given off is ^{18}O. Therefore all the O_2 by-product comes from water and not CO_2.
To take account of this, the **overall equation** for photosynthesis must be:

$$6CO_2 + 12H_2O \xrightarrow[\text{chlorophyll}]{\text{sunlight energy}} C_6H_{12}O_6 + 6O_2 + 6H_2O$$

glucose
(storing sun-energy)

This proves the need for water in photosynthesis – impossible to prove in any other way.

Fate of glucose:
(a) Converted to sucrose – for *transport* elsewhere.
(b) Converted to starch – for *storage* in leaf (transported away as sucrose by night). (Basis for leaf starch-test.)
(c) Used in *respiration*, or *amino-acid synthesis* (see unit 5.7), or fat synthesis.
(d) Converted to cellulose – making *cell walls* at growing points.

Fate of oxygen:
Diffuses out of leaves to air or water surrounding them.
Evidence that it is oxygen: see Fig. 5.1.

Fig. 5.1. Collecting and testing the gas from a pond-weed.

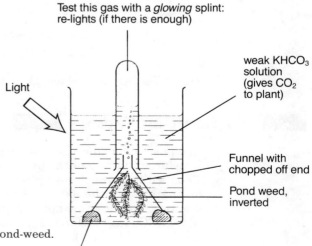

Test this gas with a *glowing* splint: re-lights (if there is enough)

Light

weak $KHCO_3$ solution (gives CO_2 to plant)

Funnel with chopped off end

Pond weed, inverted

One of 3 plasticine supports

5.2 Factors Necessary for Photosynthesis

Evidence

Plant must be *de-starched* before any experiment by keeping it in the dark for 48 hours. A leaf must now be tested for starch (as a control). The presence of starch in the leaves at the end of the experiment is evidence of photosynthesis.

1 The starch test for leaves (Fig. 5.2)

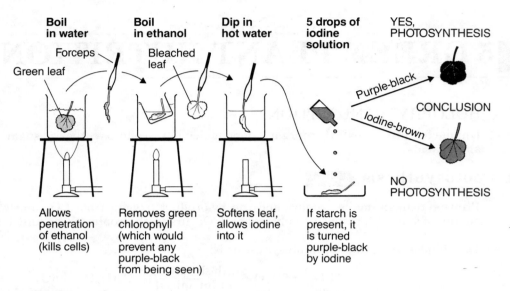

Fig. 5.2 Testing leaves for starch

2 Test the need for: ① Sunlight. ② Carbon dioxide. ③ Chlorophyll (Fig. 5.3).

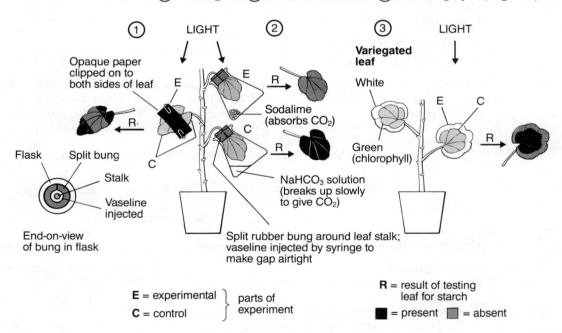

Fig 5.3 Testing the need for ① sunlight, ② carbon dioxide, ③ chlorophyll in photosynthesis

5.3 Limiting Factors

In a physiological process (such as photosynthesis) any factor which is in short supply, so that it reduces the rate of the process from its possible maximum, is said to be the limiting factor. Thus with plants photosynthesizing outdoors, *light* is limiting at dusk; *carbon dioxide* (CO_2) during most of the day; *water* probably never. *Temperature* can also be limiting (too cold – reactions too slow; too hot – destroys enzymes). *Factors closing stomata* are limiting by reducing flow of CO_2 into leaf. *Lack of magnesium* (*Mg*) in soil limits the amount of chlorophyll made in the leaf (see Fig. 5.4).

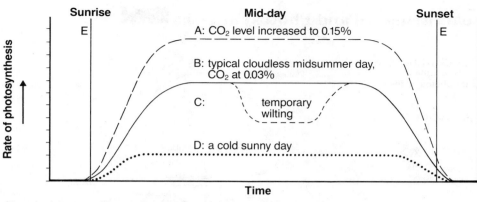

At each plateau on the graphs, a factor is limiting photosynthesis.
A and B: probably CO_2 in air
C: CO_2 reaching chloroplasts

D: temperature
E: at dawn and dusk: light

Fig. 5.4 Limiting factors for photosynthesis

5.4 Rate of Photosynthesis

Rate of photosynthesis in a water plant, e.g. *Elodea*, can be estimated by counting the *number of bubbles* per unit time coming from a cut stem. Alternatively, trap bubbles and measure *volume* per unit time in a capillary tube (see Fig. 5.5).

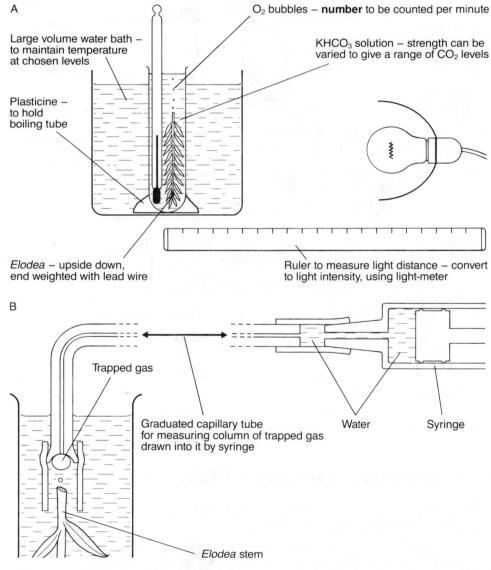

Fig. 5.5 Measuring the rate of photosynthesis in a water plant, A – by counting the number of bubbles released per minute, B – by measuring the volume of gas evolved per minute

5.5 Leaf Structure and Photosynthesis

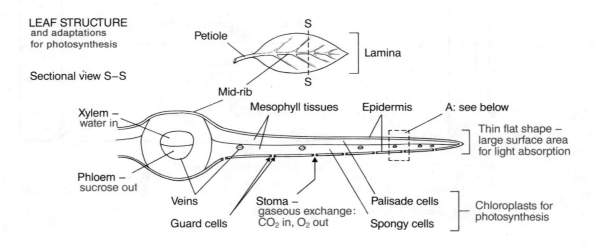

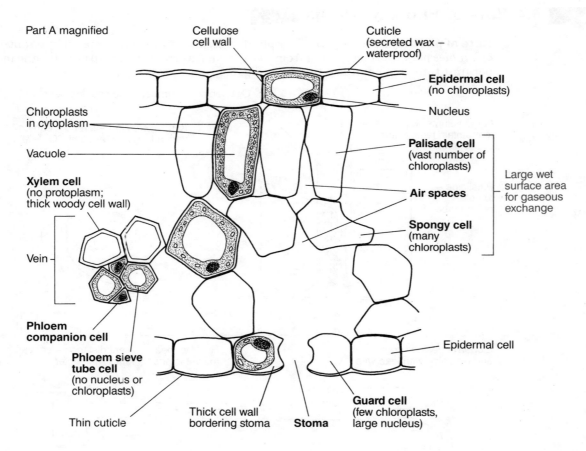

Fig. 5.6 Leaf structure

5.6 Gaseous Exchange in Leaves

By night, leaves only respire: CO_2 out, O_2 in.
By day, they photosynthesize: CO_2 in, O_2 out.
However, they also respire. But more sugars are made (by photosynthesis) than are broken down (by respiration). At dawn and dusk (very little light) the two processes break even – the *compensation point.* This point can be determined by using the apparatus below and changing the distance of the light from the plant until the indicator remains orange – as in a corked control tube (see Fig. 5.7).

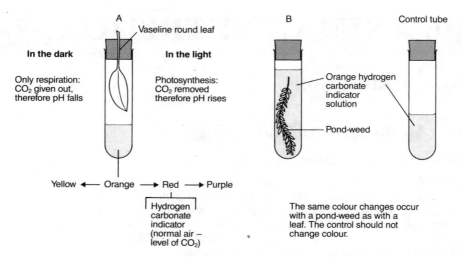

Fig. 5.7 Determining gaseous exchange: A in a leaf, B in a water plant

5.7 Amino-Acid Synthesis

Dependent on photosynthesis.

Nitrates combine with sugar products to form amino-acids.
Green plants alone can do this, at root and shoot tips (growing regions).
Amino-acids are converted (see unit 6.2) to form protein.

5.8 Mineral Salt Uptake by Roots

Absorption of salts
Mainly at root tips.
Partly at root hair region (see Fig. 16.1).
Mainly by active transport and thus oxygen is needed. Partly by diffusion (see unit 7.2).
Quite independent of water uptake by osmosis (see unit 7.3).

Evidence for need for salts
Plants are grown with roots in salt-solutions ('water culture').
Control solution contains all salts needed (see Table 4.3).
Test solutions each omit one element, e.g. – N = omit nitrates; – S = omit sulphates.
Solutions aerated to allow efficient salt uptake.

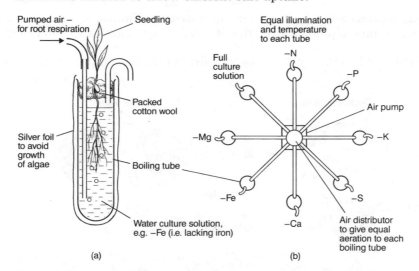

Fig. 5.8 Water culture experiment to determine the mineral salt requirements of a plant:
(a) side view of one tube, (b) plan view of experiment

In the experiment in Fig. 5.8, growth of test plants can be compared against control plants: harvest them, dry in oven at 110°C, weigh. To avoid the possibility that some seedlings grow more vigorously than others because of *genetic* differences, the plants should be from the same clone, e.g. cuttings off the same plant.

6 ANIMAL NUTRITION

HOLOZOIC NUTRITION

6.1 Feeding Methods of Animals

Animals obtain food in one of three ways:

1 As solids:
food-organisms that have to be chewed (Fig. 6.1) small enough to be ingested.
Herbivores – eat plants
Carnivores – eat animals
Omnivores – eat plants and animals

2 As solids in suspension: tiny food-organisms in water that must be strained out of it – plankton (plants and animals)
Filter-feeders, e.g. mosquito larva (Fig. 21.18)

3 As liquids:
(a) juices extracted from living hosts, without killing them.
Parasites (see unit 19.3)
(b) liquid nutriment produced by digesting dead food externally and then sucking it up
Saprozoites, e.g. house-fly (Fig. 6.1)

Adaptations necessary for each feeding method

1 Herbivores: food does not run away, but large quantities must be gathered since food is relatively poor in quality. Herbivores include locusts, snails, deer and sheep.

2 Carnivores: have to capture and overcome prey, e.g. by cunning (dogs), traps (spiders' webs), poisons (cobras) and sharp weapons (claws, teeth). (See Fig. 6.6.)

3 Omnivores: adaptations for feeding are intermediate between those of herbivores and carnivores, e.g. human teeth. Often very successful animals since they vary their food according to availability, e.g. cockroaches, rats, pigs and Man.

4 Filter feeders: require sieves, e.g. *Daphnia*, Fig. 21.10. Baleen whales trap 'krill' (shrimps) on frayed edges of whale-bone plates hanging down in mouth cavity, open to the sea as they swim.

5 Parasites: endo-parasites bathe in nutritious liquids, e.g. blood or digested food in gut of host, absorbing food directly through 'skin' – no gut, e.g. *Trypanosoma*, tape-worm (see unit 21.15).

Ecto-parasites pierce their host to suck out nutritious liquids, e.g. mosquito, flea (blood). Aphids (greenfly) pierce phloem: pressure of sap forces it into insect (and some out as 'honey-dew').

6 Saprozoites: need no jaws, only tubes for saliva (down) and liquid food (up), with pumps (see Fig. 6.1.)

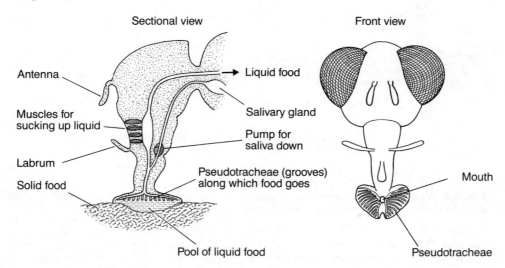

Fig 6.1 Mouthparts of a house-fly – for sucking liquid food (digested externally)

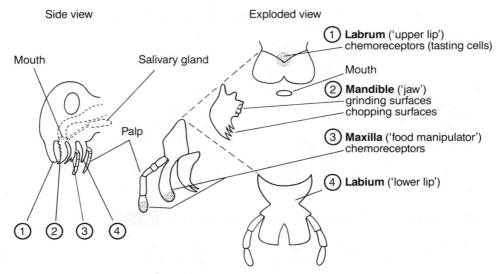

Fig. 6.2 Mouthparts of a chewing insect, e.g. locust (herbivore), cockroach (omnivore), ground beetle (carnivore)

6.2 Digestion and its Consequences

All animals **ingest** food via the mouth into a gut (or equivalents). Exceptions: parasites which bathe in food. In the gut, food is **digested** in two ways:

(a) *physically* – by chewing or grinding (important in herbivores), stomach churning and peristalsis. This increases the surface area of food, making it easier for (b) below.

(b) *chemically* – by enzymes (see unit 1.5) which hydrolyse large molecules into their small basic units (see unit 4.4). Without this, large insoluble food molecules would not be small enough to be **absorbed** through the membranes of gut cells: e.g.

$$\text{starch} + \text{water} \xrightarrow[\text{enzymes}]{\text{carbohydrase}} \text{monosaccharides}$$

$$\text{fat} + \text{water} \xrightarrow[\text{enzymes}]{\text{lipase}} \text{fatty acids} + \text{glycerol}$$

$$\text{protein} + \text{water} \xrightarrow[\text{enzymes}]{\text{protease}} \text{amino-acids}$$

these molecules are now soluble and small enough for absorption

Absorbed food is then **assimilated** (used or stored) into the body. Storage occurs when enzymes condense the small units of foods into large molecules (reverse of hydrolysis). For example:

$$\text{amino-acids} \xrightarrow[\text{enzymes}]{\text{condensing}} \text{protein} + \text{water}$$

Indigestible food is **egested** (eliminated) through the anus or equivalent. Most animals have no enzymes to digest cellulose – hence special adaptations of herbivores. (See unit 6.8.) Mammal *faeces* include egested roughage, bacteria, mucus, dead cells and water, and excreted bile pigments. (See Fig. 6.13.)

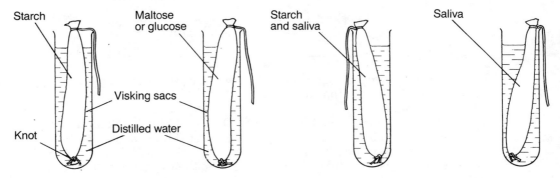

Fig. 6.3 A model of digestion and absorption

The 'gut' is Visking membrane, known to allow small molecules (sugar) but not large ones (starch) through its pores.

The enzyme in saliva turns starch into maltose. Maltose is turned into glucose in the gut (see 'villi', Fig. 6.10).

Four sacs of Visking containing different solutions are placed in distilled water. After 30 minutes the water is tested for starch and for reducing sugar. (The saliva must not contain sugar from sweets.)

Results of tests on water:

Benedict's test (Table 4.4)	–	+	+	–
Iodine test (Table 4.4)	–	–	–	–

Conclusions: **1** Sugars, but not starch, pass through Visking pores.
                  2 Saliva turns starch into sugar

6.3 Experiments with Digestive Enzymes

Each enzyme works best at a certain temperature and pH (these are its 'optimum' conditions). Outside these conditions enzymes may cease to work or may even be destroyed.

Example 1 Investigating the effect of temperature on digestion of starch by salivary amylase.

Method:
1 Add 5 cm³ of 1% starch solution to each of 5 boiling tubes and 1 cm³ of saliva diluted with water to 4 test tubes as shown in Fig. 6.4(a).

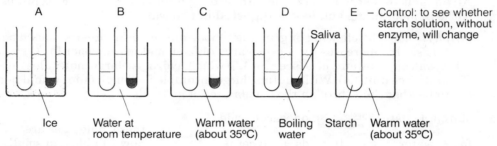

E – Control: to see whether starch solution, without enzyme, will change

Saliva

| Ice | Water at room temperature | Warm water (about 35°C) | Boiling water | Starch | Warm water (about 35°C) |

Fig. 6.4(a)

2 Leave the starch and the enzyme for at least two minutes, to gain the temperature of the water bath.
3 Pour the saliva into the boiling tube next to it, so mixing it with the starch. Note the time immediately.
4 Using a separate dropper for each tube, test one drop from each boiling tube with iodine, as shown in Fig. 6.4(b).

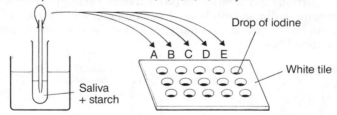

One drop from each of tubes A, B, C, D, E, every 30 seconds

Drop of iodine

A B C D E

White tile

Saliva + starch

Fig. 6.4(b)

5 Note the time when each drop no longer turns the iodine blue-black (i.e. starch is digested). Do not test for longer than 15 minutes.
Possible results:
A – still blue-black after 15 minutes   D – still blue-black after 15 minutes
B – changes to brown at 8 minutes   E – still blue-black after 15 minutes
C – changes to brown at 2 minutes
6 Now put the boiling tubes from A and D into the warm water bath C and test them with iodine after 5 minutes (once only).
Results:
A – brown colour   D – blue-black
Conclusions:
1 Digestion proceeds faster at warm temperatures than at cold (A, B, C).
2 At low temperatures, the enzyme is inactive but not destroyed (A, step **6**).
3 At water's boiling point, the enzyme is destroyed (D, step **6**).

Example 2 Investigating the effect of pH on digestion of egg albumen (protein) by pepsin.

Method:
1 Put in each of 6 tubes a 5 mm cube of cooked egg white and a thymol crystal (to prevent bacteria digesting the egg). Then add 2 cm³ of 0.1M solutions to affect the pH as shown in Fig. 6.5.

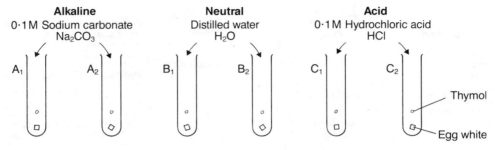

Fig. 6.5

2 Add 2 cm³ pepsin solution to A_1, B_1 and C_1, but not to A_2, B_2 and C_2 (which are controls used to see whether Na_2CO_3, water and HCl alone digest egg white).
3 Incubate the tubes in a warm place (about 35°C) for 24 hours and then look at the cubes.

Results:

Not digested:	Slightly digested:	Totally digested:
in tubes A_1, A_2, B_2, C_2	in tube B_1	in tube C_1
Sharp edges	Smaller cube with fuzzy edges	Cube absent

Conclusion: pepsin requires acid conditions to digest cooked albumen.

6.4 Mammal Teeth

Mammals are the only vertebrate group with *differentiated* teeth (four types with special uses):
1 **Incisors** – for biting off food
2 **Canines** – for stabbing, holding prey
3 **Pre-molars** – for grinding
4 **Molars** – for grinding

First set of teeth are shed ('milk teeth'): 20, made up of 8 **I**, 4 **C**, 8 **Pm** in Man.
Adult set includes 'wisdoms' (back molars): 32, made up of 20 larger replacements and 12 **M**.
Structure of teeth: layers of modified bone nourished from pulp cavity and shaped according to function. (See Fig. 6.6)

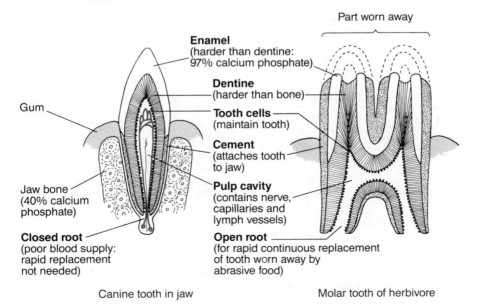

Fig. 6.6 Vertical section through two kinds of teeth

6.5 Mammal Alimentary Canal

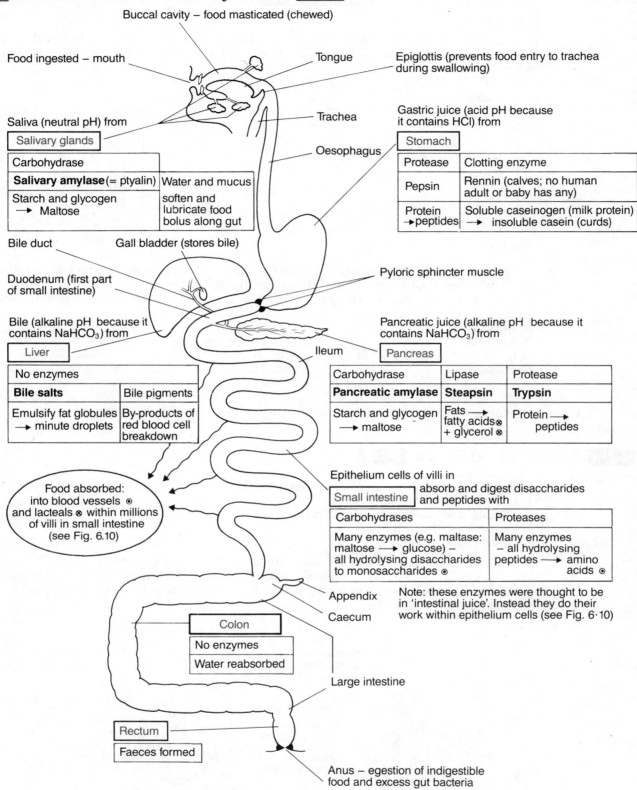

Buccal cavity – food masticated (chewed)

Food ingested – mouth

Tongue

Epiglottis (prevents food entry to trachea during swallowing)

Trachea

Oesophagus

Saliva (neutral pH) from

Salivary glands	
Carbohydrase	
Salivary amylase (= ptyalin)	Water and mucus
Starch and glycogen → Maltose	soften and lubricate food bolus along gut

Gastric juice (acid pH because it contains HCl) from

Stomach	
Protease	Clotting enzyme
Pepsin	Rennin (calves; no human adult or baby has any)
Protein → peptides	Soluble caseinogen (milk protein) → insoluble casein (curds)

Bile duct

Gall bladder (stores bile)

Duodenum (first part of small intestine)

Pyloric sphincter muscle

Bile (alkaline pH because it contains $NaHCO_3$) from

Ileum

Pancreatic juice (alkaline pH because it contains $NaHCO_3$) from

Liver	
No enzymes	
Bile salts	Bile pigments
Emulsify fat globules → minute droplets	By-products of red blood cell breakdown

Pancreas		
Carbohydrase	Lipase	Protease
Pancreatic amylase	**Steapsin**	**Trypsin**
Starch and glycogen → maltose	Fats → fatty acids⊗ + glycerol ⊗	Protein → peptides

Food absorbed: into blood vessels ⊙ and lacteals ⊗ within millions of villi in small intestine (see Fig. 6.10)

Epithelium cells of villi in

Small intestine	absorb and digest disaccharides and peptides with
Carbohydrases	Proteases
Many enzymes (e.g. maltase: maltose → glucose) – all hydrolysing disaccharides to monosaccharides ⊙	Many enzymes – all hydrolysing peptides → amino acids ⊙

Appendix

Caecum

Note: these enzymes were thought to be in 'intestinal juice'. Instead they do their work within epithelium cells (see Fig. 6·10)

Colon	
No enzymes	
Water reabsorbed	

Large intestine

Rectum

Faeces formed

Anus – egestion of indigestible food and excess gut bacteria

Fig. 6.7 Treatment of food from mouth to anus in mammals (based on Man)

Abnormal passage of food:

1 Vomiting: strong contraction of stomach fountains food, containing toxins, too much salt etc., out of mouth.

2 Diarrhoea: irritation of villi causes too much mucus and intestinal juice secretion. Sweeps out harmful microorganisms, eg. cholera in liquid faeces.

3 Constipation: waste can be solidified too much by colon: hard faeces can cause bleeding. Roughage (fibre) in diet helps prevent this.

6.6 Dental Health

1 Growing healthy teeth require:
(*a*) food rich in *calcium* (unit 4.3) and *vitamin D* to help in its absorption (unit 4.5);
(*b*) *fluoride* from fluoridated water or toothpaste to harden enamel.

2 Maintaining healthy teeth requires controlling the bacteria around teeth by *dental hygiene*. There are two main dental diseases.
(*a*) *Caries* (holes in teeth) results from bacteria turning sugars into acids. Acids dissolve enamel, allowing bacteria to rot dentine.
(*b*) *Periodontal disease* (teeth fall out) results from bacteria entering space between tooth and unhealthy gums. They rot fibres holding teeth in socket.

Dental Hygiene
(*a*) *Rinse mouth with water* to remove sugars, after meals (or sweets).·
(*b*) *Brush teeth* with fluoride toothpaste, especially before sleep, to remove food particles and bacteria. These form 'plaque' – a coating stuck to teeth – if left. Disclosing tablets stain it red.
(*c*) *Massage gums* by eating crisp foods; and as part of tooth-brushing to keep them healthy. This prevents exposure of neck and root of tooth to bacteria.
(*d*) *Orthodontic treatment* improves hygiene by uncrowding teeth, making them easier to keep clean.

6.7 Herbivores and Carnivores: Teeth and Jaws

Carnivores need only to swallow lumps of food; protease enzymes can do the rest.
Herbivores must grind food to give a large surface area for cellulases to act upon (unit 6.8).

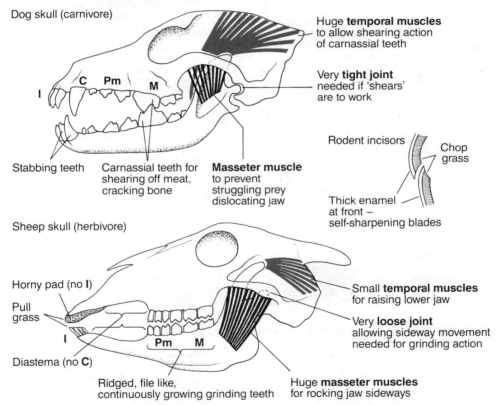

Fig. 6.8 Comparison of herbivore and carnivore jaws and teeth

6.8 Herbivores and Carnivores: the Gut

Carnivore: gut *short* – food is largely protein so it is easy to digest.

Herbivore: gut *long* – no mammal produces cellulases (cellulose-digesting enzymes). Aid comes from bacteria which do have *cellulases*. These live in symbiosis (see unit 19.3) in *rumen* of ruminants, e.g. cows, sheep; or *caecum*, e.g. of horses, rabbits. Rabbits eat their green nutritious faeces from first passage through gut ('refection'), absorbing more food during second passage. Horses do not refect (see Fig. 6.9), and can absorb food through the caecum.

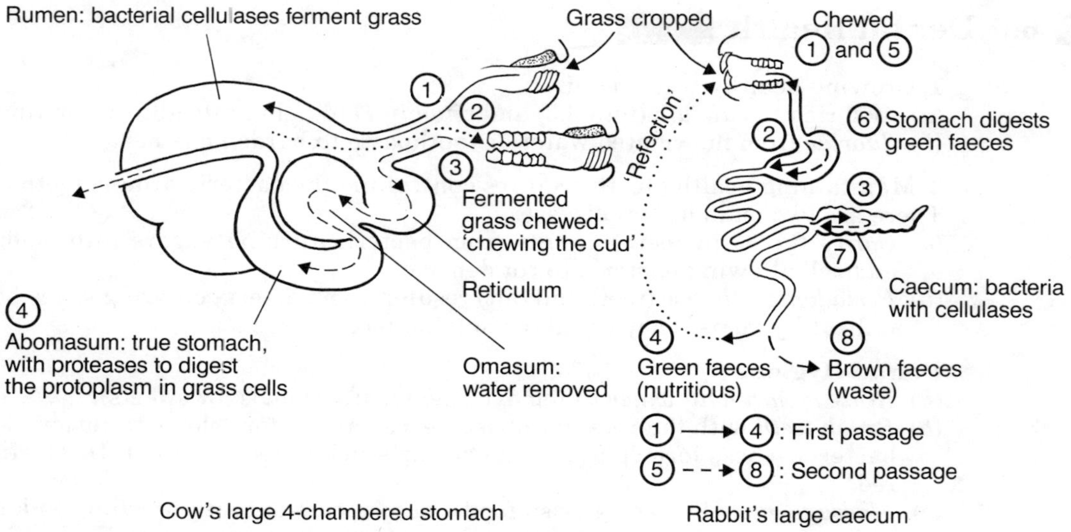

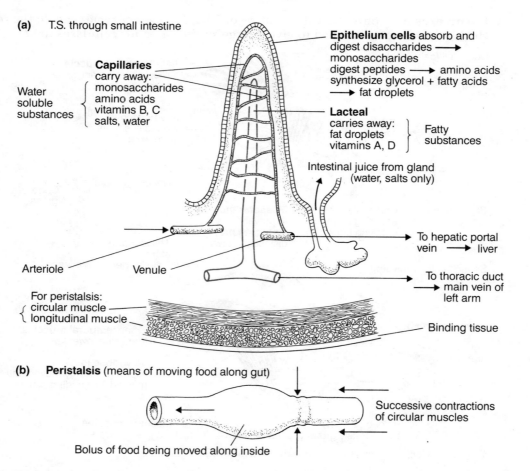

Rumen: bacterial cellulases ferment grass

Grass cropped

Chewed ① and ⑤

⑥ Stomach digests green faeces

Fermented grass chewed: 'chewing the cud'

Reticulum

'Refection'

④ Abomasum: true stomach, with proteases to digest the protoplasm in grass cells

Omasum: water removed

Caecum: bacteria with cellulases

④ Green faeces (nutritious) → ⑧ Brown faeces (waste)

① ⟶ ④ : First passage
⑤ --→ ⑧ : Second passage

Cow's large 4-chambered stomach

Rabbit's large caecum

Fig. 6.9   Adaptations to herbivorous diet

6.9   Absorption of Food at a Villus

(a)   T.S. through small intestine

Epithelium cells absorb and digest disaccharides ⟶ monosaccharides digest peptides ⟶ amino acids synthesize glycerol + fatty acids ⟶ fat droplets

Capillaries carry away: monosaccharides amino acids vitamins B, C salts, water

Water soluble substances

Lacteal carries away: fat droplets vitamins A, D } Fatty substances

Intestinal juice from gland (water, salts only)

To hepatic portal vein ⟶ liver

Arteriole   Venule

To thoracic duct ⟶ main vein of left arm

For peristalsis: { circular muscle longitudinal muscle

Binding tissue

(b)   **Peristalsis** (means of moving food along gut)

Successive contractions of circular muscles

Bolus of food being moved along inside

Fig 6.10   (a) Enlarged longitudinal section of a villus (millions lining the small intestine) (b) Peristalsis

6.10   Storage of Food

1   Monosaccharides, e.g. glucose: turned into glycogen for storage in liver and muscles; excess converted to fats stored under skin.
2   Fatty substances: stored in liver (including vitamins A, D) and under skin.
3   Amino-acids: used immediately in growth and repair. *Not* stored; excess deaminated in liver.

6.11 The Liver

A large organ, concerned with homeostasis by
metabolizing food and poisons and removing
unwanted cells. Stores foods and blood. Receives
blood from two sources (Fig. 6.11); discharges bile.

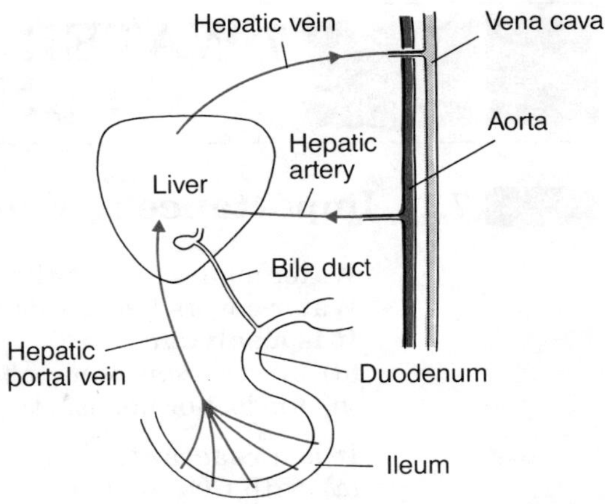

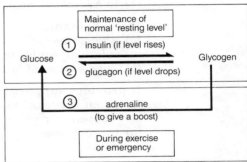

Fig. 6.11 The liver and its blood supply

1 Stores *glucose* as glycogen, turning it back to
glucose when needed. This is under the control of
three hormones which keep the *blood level* of
glucose constant.

Fig. 6.12 An example of homeostasis (see unit 10.9)

2 Deaminates *excess amino-acids* to give two parts:
(*a*) nitrogen-containing part (amine) becomes urea – excreted by kidneys;
(*b*) remainder (the acid) can be respired to give energy.
3 Stores *iron* from worn-out red blood cells, which it breaks down, excreting *bile pigments* in
the process.
4 Makes *poisons* harmless, e.g. ethanol drunk or toxins from gut bacteria.
5 Makes bile salts which emulsify *fats* in the intestine.

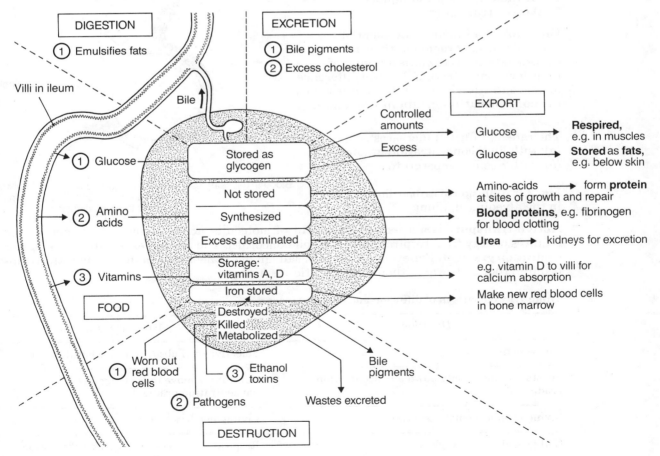

Fig. 6.13 The liver's five roles

7 WATER UPTAKE AND LOSS IN PLANTS AND ANIMALS

7.1 Importance of Water

Water makes up two-thirds or more of living active cells.
Water covers two-thirds of the globe – a very important habitat for organisms.

It is a solvent:
(*a*) all *reactions* of metabolism occur in solution.
(*b*) foods, hormones, etc. are *transported* in solution (in blood, sap).

It is a reactant:
(*a*) with CO_2 during *photosynthesis*.
(*b*) in *hydrolysis* reactions, e.g. digestion.

It is a coolant:
(*a*) *absorbs a lot of heat* without much change in temperature, thus keeping habitats like the sea relatively stable in temperature.
(*b*) *removes a lot of heat* when evaporated, keeping bodies cool, e.g. in sweating, transpiration.

It provides support:
(*a*) aquatic organisms need less strong skeletons than land organisms because water's *'buoyancy effect'* (*Archimedes force*) makes them 'lighter'.
(*b*) turgor pressure in plant cells supports leaves and herbaceous plant stems; without it they wilt.

It is a lubricant:
E.g. synovial fluid in joints (see unit 13.8); mucus in guts.

7.2 Diffusion and Active Transport

Substances move into cells by:

1 Diffusion (gases and liquids)
2 Active transport

Diffusion is a random movement of molecules from a region of their high concentration (A) towards a region of their low concentration (B). The difference in concentration between A and B is called the **concentration gradient** (see Fig. 7.1).

The rate of diffusion increases:
(*a*) with high concentration gradients
(*b*) with rise in temperature

CO_2 enters at 6 kPa

CO_2 enters at almost 0 kPa (0.03% of air)

Air in alveolus

Blood in capillary

Diffusion gradient is 6 → 0 between blood and air. So CO_2 moves from blood to air by diffusion

Fig. 7.1 Diffusion of carbon dioxide (CO_2) in the alveolus of a lung

Active transport is a selective movement of molecules across living cell membranes. This requires energy from respiration in the cell concerned. The molecules usually move *up* a concentration gradient. For example, mineral salts in *low* concentration in soil may still be absorbed into root cells where their concentration is *higher*.

Table 7.1 Comparison of diffusion and active transport

Diffusion	*Active transport*
Not selective	Selective (cell absorbs only what it needs)
Substances move only down a concentration gradient	Substances move in even *against* a concentration gradient
Living membrane not essential	Living membrane essential
Cell provides no energy	Respiration provides energy for absorption

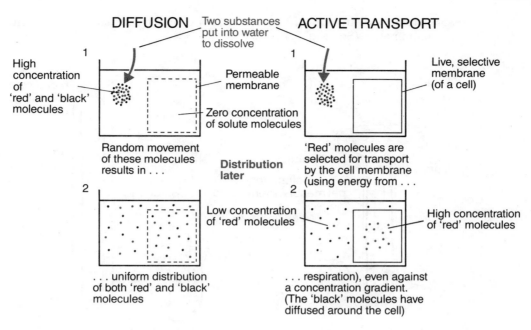

Fig. 7.2   Diffusion contrasted with active transport

When diffusion plays a large part in a biological process the organs concerned, e.g. leaf, lung, have a large surface area (see unit 22.6).

7.3 Osmosis

Osmosis is the diffusion of water *only*, through a selectively permeable membrane, from where *water is in high concentration* (a weak solution) to where *water* is in low concentration (a strong solution). (See Fig. 7.3.)

Requires *no* respiration (cf. active uptake of salts in roots).

Requires *live* cell membrane for osmosis to occur in cells, but will happen with suitable non-living membranes (e.g. Visking dialysing membrane).

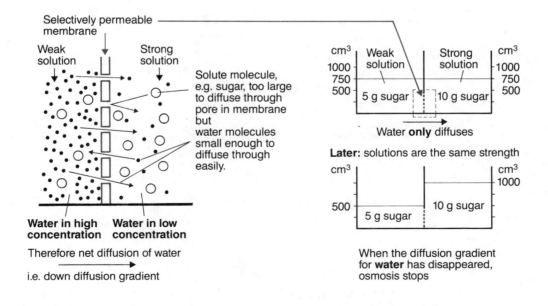

Fig. 7.3 Osmosis

All solutions have an **osmotic potential**, which can be measured as an **osmotic pressure** by using a manometer. (See Fig. 7.4.)

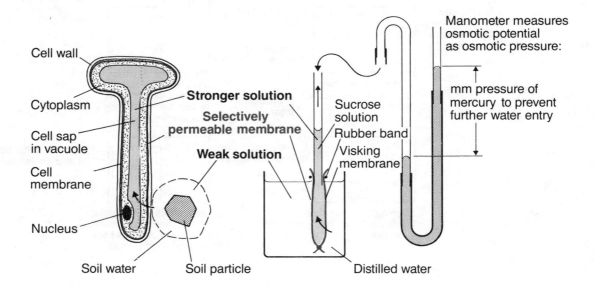

Fig. 7.4 Comparison of osmosis in a living cell (root hair) and a non-living system

Cells prevent continued flow of water into them (which would burst them) by **osmoregulating** (see unit 7.4 and Fig. 7.7).

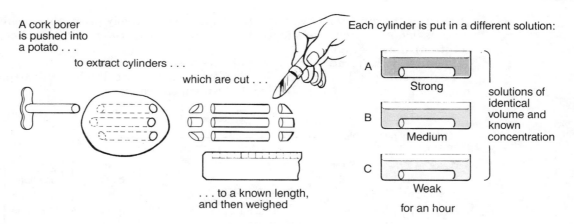

The cylinders are removed, dried at their ends and rolled gently on blotting paper. They are then weighed again and measured for length. These measurements are put on graphs

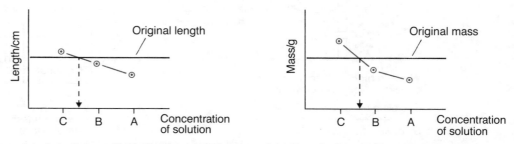

At the point where the graph cuts the line indicating the *original* length or mass the average osmotic potential of potato cells may be read off (see ↓). If this strength of solution were to be prepared, the potato cylinders should neither increase nor decrease in length or mass, i.e. no osmosis would occur

Fig. 7.5 Experiment to determine the osmotic potential of potato tissue

7.4 Osmosis in Cells

Plant cells

Cells, in nature, fluctuate between being flaccid and fully turgid. However, plasmolysis is relatively rare (except in experiments) and will result in the cell's death if it is prolonged, e.g. when the cell suffers prolonged drying (see Fig. 7.6).

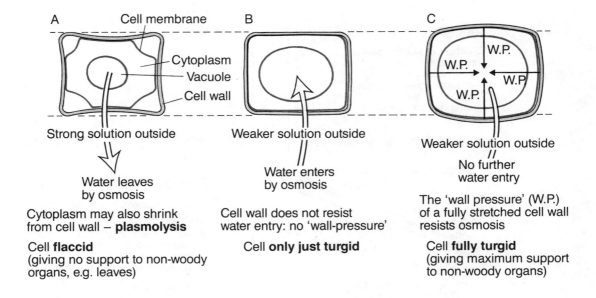

Fig. 7.6 Osmosis in plant cells

Such changes may be seen down a microscope when strips of rhubarb epidermis are mounted in strong, medium and weak solutions on three different slides.

Animal cells

Amoeba takes an active part in ejecting water gained by osmosis from its weak solution (fresh water) habitat, to prevent bursting (see Fig. 7.12).

Red blood cells will burst in fresh water and shrink in strong solutions. They rely on kidneys to keep the plasma at the right concentration (see unit 10.2).

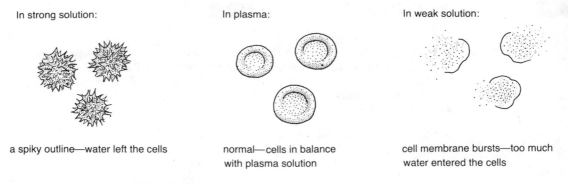

Fig. 7.7 Osmosis in red blood cells

7.5   Water Uptake and Loss in Flowering Plants

1   Leaves and green stems are waterproofed by a waxy *cuticle*, but most keep open *stomata* to get CO_2 for photosynthesis. Through stomata, **transpiration** (the loss of water vapour via the aerial parts of a plant) occurs. This creates a *suction upward* of water from below.

2   Old stems and roots are waterproofed with cork (of bark). Their xylem allows passage of water. Some water-loss occurs via *lenticels* (pores in bark).

3   Young roots – particularly *root-hair* region – absorb water by osmosis. This continues owing to suction generated by transpiration.

If soil water supply dries up, leaf cells become flaccid and leaf *wilts*. Only *after* this will guard cells become flaccid, closing stomata, thus conserving water but also stopping photosynthesis (see unit 5.3 and Fig. 7.8).

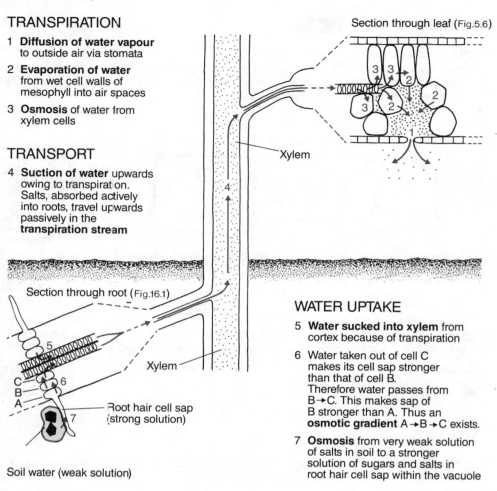

TRANSPIRATION

1 **Diffusion of water vapour** to outside air via stomata

2 **Evaporation of water** from wet cell walls of mesophyll into air spaces

3 **Osmosis** of water from xylem cells

TRANSPORT

4 **Suction of water** upwards owing to transpiration. Salts, absorbed actively into roots, travel upwards passively in the **transpiration stream**

Section through leaf (Fig.5.6)

Xylem

Section through root (Fig.16.1)

Xylem

Root hair cell sap (strong solution)

Soil water (weak solution)

WATER UPTAKE

5 **Water sucked into xylem** from cortex because of transpiration

6 Water taken out of cell C makes its cell sap stronger than that of cell B. Therefore water passes from B→C. This makes sap of B stronger than A. Thus an **osmotic gradient** A→B→C exists.

7 **Osmosis** from very weak solution of salts in soil to a stronger solution of sugars and salts in root hair cell sap within the vacuole

Fig. 7.8 Water uptake, transport and loss in a flowering plant

7.6 Guard Cells and Stomata

Guard cells are kidney-shaped green cells found in pairs in the epidermis ('skin') of leaves and green stems. The pore between them is a **stoma** (plural: **stomata**). This appears when guard cells are turgid; disappears when they are flaccid (see Table 7.2).

The flaccid cells (red outline) take in water by osmosis from epidermis cells. This stretches the thin outer walls of guard cells, so bending the thickened inner walls (black outline) to open up a pore between them.

Table 7.2. Features of guard cells in the turgid and flaccid states.

	Stoma open	Surface view of two guard cells in the turgid and flaccid states	Stoma closed
Osmotic potential of cell sap	High		Low
Turgidity	Turgid	Turgid cells / Flaccid cells; Open stoma / Closed stoma	Flaccid
Gas exchange and transpiration	Possible		Impossible
Normal rhythm	Open in day		Closed at night

Determining whether conditions around the leaf affect the opening and closing of stomata:

A film of nail varnish or stencil-correcting fluid painted onto a non-hairy leaf can be peeled off with forceps when dry (30 sec). Under the microscope the dried film bears impressions of stomata. The number open can be recorded, e.g. '5/20' for the 20 stomata observed.

Films from the *same* plant leaf after being in different conditions (e.g. of light, dark, CO_2, wind and temperature) can be compared. *One* condition should be changed at a time.

Different species of plant can behave differently.

7.7 Transpiration

Transpiration is the loss of water vapour through the aerial parts of a plant. It occurs
 (i)   mainly through open stomata
 (ii) through waxy cuticle (a small amount)
Functions:
 (i)   provides a means of transporting salts upward in xylem
 (ii) cools the leaf heated by the sun by evaporation (cf. sweating).

Factors raising transpiration rate (opposite conditions lower the rate)

1   High temperature – provides more energy to evaporate water.

2   Low humidity – greater diffusion gradient between air inside leaf spaces and the drier air outside.

3   Open stomata – thousands of pores per leaf (usually open in *sunlight*).

4   Wind – removes water molecules as fast as they arrive outside stomata, thereby maintaining high diffusion rate. Water vapour 'pumped out' due to bending and unbending of leaf. (Severe buffeting by wind actually closes stomata, reducing transpiration.)

Measurement of transpiration rate
(Temperature, humidity, and wind must be recorded.)

1   Weighing – a leaf, or cut shoot, in a test-tube of water covered by oil; or a whole pot plant, the pot and soil sealed off in a polythene bag.

2   Cobalt chloride – blue when anhydrous (dry), turns pink when hydrated (moist). Thus dry blue cobalt chloride paper, sellotaped to upper and lower leaf surfaces, green stems and bark-covered stems turns pink with moisture of transpiration. Timing how long it takes compares rates.

3   Potometer – measures water uptake (not loss) of a cut shoot (a little of the water is used in photosynthesis). Change *one* condition at a time to determine which factor has greatest effect.

Note: light and dark affect opening and closing of stomata. Light may also have a heating effect.

Allow time for plant to adjust to new conditions before taking new measurement of rate.

Never allow air to get into cut end of shoot (air bubbles block the xylem) – cut shoot under water, and keep the cut wet (see Fig. 7.9).

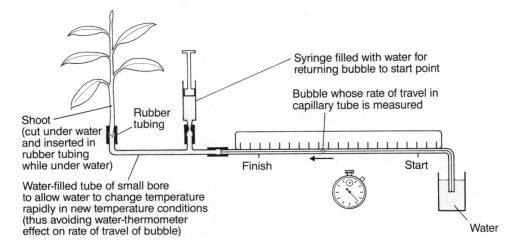

Fig. 7.9 A potometer in action

Evidence for pathway of water in a stem
If cross sections of a stem are taken 30 minutes after putting a leafy shoot in re['only the] xylem is stained red (see Fig. 7.10). The xylem of a stem in water, used as a co[', does not] become red.

7.8 Transport of Organic Food

Flows through **phloem** sieve tube cells in bark (see Fig. 7.11). Flow rate is affected by temperature, available oxygen, poisons – suggests a mechanism involving *living* cells. Mechanism not fully understood.

Flows both *upwards and downwards*. Photosynthesized sugars are transported as sucrose (see unit 4.4) from leaves *up* to stem tips (for growth); to fruits and seeds (for storage as starch); *down* to root tips (for growth); and to or from storage organs, e.g. tubers (see unit 14.2).

Evidence for pathway of organic food

Ring barking: sugars accumulate where bark ends (due to cutting).

Tracers: radioactive $^{14}CO_2$ supplied to a photosynthesizing leaf becomes part of sucrose (or other organic molecules). Cross-sections of stems below such leaves, when placed next to photographic film (for a week in a refrigerator), will become exposed only where there is phloem (see Fig. 7.10). Control film remains unexposed. This shows that radioactive sucrose is transported in phloem.

Systemic insecticides when sprayed onto leaves are absorbed by them and pass to the phloem. Insects, e.g. aphids, sucking out the sugary sap, are thus poisoned.

7.9 Tissues in the Stem and Root

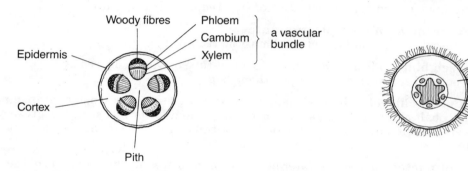

Section of young stem

Epidermis · Woody fibres · Phloem · Cambium · Xylem · a vascular bundle · Cortex · Pith

Section of young root

Root hairs · Cortex · Phloem · Xylem

Fig. 7.10 Cross sections of young stem and root of a flowering plant

Epidermis: waterproof outer 'skin' of waxed non-green cells – and some guard cells (see leaf: Fig. 5.6).

Root hairs: water-absorbing cells having a large surface area (see Fig. 7.4).

Cortex and pith: large cells capable of storing food, e.g. starch
Cambium: cells that can divide to cause growth in diameter
Woody fibres: give strength in the wind
Xylem vessels (also woody): (*a*) give strength
(*b*) transport water and mineral salts *upward*
Phloem sieve tubes: transport sugars and amino-acids *up and down*
(These cells/tissues are common to both stem and root)

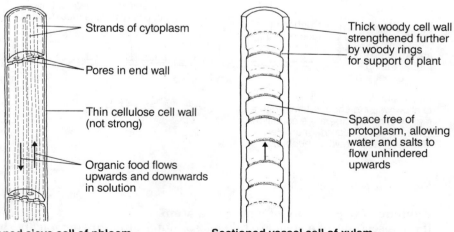

Strands of cytoplasm

Pores in end wall

Thin cellulose cell wall (not strong)

Organic food flows upwards and downwards in solution

Thick woody cell wall strengthened further by woody rings for support of plant

Space free of protoplasm, allowing water and salts to flow unhindered upwards

Sectioned sieve cell of phloem

Sectioned vessel cell of xylem

Fig 7.11 The structure of cells conducting food in xylem and phloem.

7.10 Water Uptake and Loss in Animals

Animals have two problems that plants do not have:

1 Lack of cell walls to prevent excess water entering (see unit 1.1). Thus cells are liable to burst (see Fig. 7.7) unless they osmoregulate by ejecting water, e.g. via contractile vacuoles or 'kidneys'.

2 Excretion of nitrogenous wastes which need water for their removal:

(*a*) **ammonia** (NH_3) – very poisonous; needs large quantities of water to dilute and remove it. Fresh-water animals, particularly, excrete this.

(*b*) **urea** ($CO(NH_2)_2$) – less poisonous; needs some water to remove it. Many terrestrial animals, e.g. mammals, excrete this.

(*c*) **uric acid** – not poisonous, since insoluble; can be removed as a paste. Essential for all animals laying eggs on land to avoid poisoning of embryo, e.g. insects, birds. Very little water wasted.

As with plants, animals have three problems: **obtaining** water, **conserving** what has been obtained and **removing excess** water that has entered. These problems and the ways they are solved differ according to the animal and the habitat in which it lives (see Fig. 7.12).

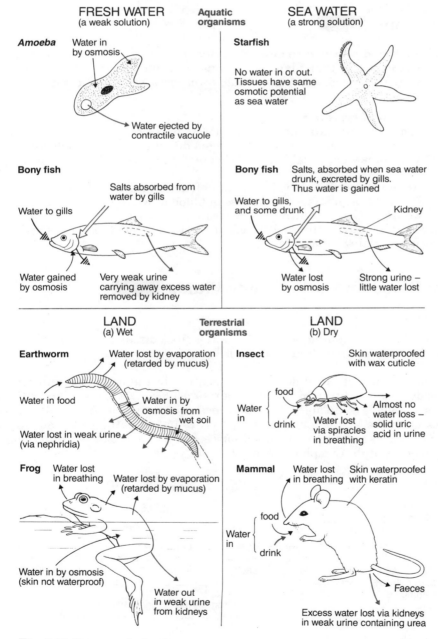

Fig. 7.12 Osmoregulation in animals:
→ problem created by animal's environment
→ corrective measures employed by animal

1 In **fresh water**, water *enters* by osmosis, tending to flood tissues since they have a higher osmotic potential than their external surroundings.
2 In **sea water**, the water inside tissues tends to *leave* by osmosis into the sea since its salty water has a higher osmotic potential than tissues in many cases.
3 In **wet-land** habitats, water still *enters* tissues by osmosis through non-waterproof skin, but there is the hazard of desiccation in the air. Such animals do not drink but can gain some water from food.
4 In **dry-land** habitats, animals must have waterproof skins to prevent desiccation in the air, replacing what they lose in breathing and excreta by *drinking*. Egg-layers excrete uric acid, so little water is lost in urine.

8 THE BLOOD AND LYMPHATIC SYSTEMS

8.1 Blood Systems

The need for blood pumped to cells by a heart
Animals are more active than plants and diffusion would be too slow to supply cells with their needs and remove their wastes. A more *rapid transport system* is necessary to prevent them from dying.

Functions of blood systems
1 Supply **foods** – sugars, fats, amino-acids, vitamins, salts, water.
2 Supply **oxygen** – (exception: insects – oxygen direct to cells at tracheoles).
3 Supply **hormones** – chemical 'messages' controlling metabolism and development (see unit 12.10).
4 Supply **leucocytes** – white blood cells for defence against invading organisms.
5 Supply **clotting materials** – to stop loss of blood at wounds.
6 Remove **wastes** – CO_2 and nitrogenous wastes, e.g. urea.
7 Carry **heat** – either away from cells, e.g. muscle, to cool them, or to cells needing to be warmed up, e.g. during 'sunning' of lizards.

8.2 Mammal Blood and Other Body Fluids

Blood consists of:

(*a*) **plasma**, a straw-coloured liquid (90% water, 10% dissolved substances);
(*b*) **cells**, a variety of kinds (see Table 8.1).

Exact composition of blood depends on location in the body (see unit 8.6) and on health. Human body has 5–6 litres of blood, about 10% of body-weight, pumped through arteries, capillaries and veins (see Table 8.2). Blood does not bathe cells. At capillaries, **tissue fluid** – a colourless nutritive liquid containing O_2 – oozes out to bathe cells and carry away wastes. Tissue fluid returns mainly into the capillaries; but the excess passes into the lymph vessels to become part of **lymph**. Lymph is discharged into veins.

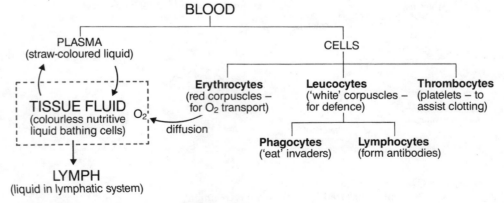

Fig. 8.1 Constituents of blood and their functions

Serum is plasma less fibrinogen (protein needed for clotting). Stored by hospitals for transfusions.

Plasma consists of:
1 Water – (90%) solvent for substances listed below; carrier of heat (for temperature regulation).
2 Blood proteins – (7%) e.g. fibrinogen (for blood clotting); antibodies (for defence against pathogens); and albumen (for osmosis, Fig. 8.8).
3 Soluble foods – (1%) e.g. glucose, oil droplets, amino-acids (from digestion).
4 Mineral salts – as ions, e.g. Na^+, Cl^-, Ca^{2+}, HCO^{-3} (hydrogencarbonate, the main method of transporting CO_2).
5 Wastes – e.g. CO_2, urea.
6 Hormones – in minute traces, e.g. adrenaline and insulin.
7 Gases – small quantities of, e.g. O_2, N_2.

Table 8.1 Blood cells and their functions

Cell Structure		No./mm³	Formation	Destruction	Function of cells
Erythrocyte in section	Bi-concave cell with no nucleus. Cytoplasm: mainly red haemoglobin	5 million (more at high altitudes)	In red bone marrow, e.g. of ribs, vertebrae	In liver – by-product: bile pigments Life: 2–3 months	1 Haemoglobin (Hb) combines with O_2 to form unstable oxyhaemoglobin ($Hb.O_2$) at lungs – Passes **oxygen to tissues** 2 O_2 detaches from Hb at capillaries, diffusing into the tissue fluid going to cells (see unit 9.8) $Hb + O_2 \xrightarrow[\text{tissues}]{\text{lungs}} Hb. O_2$
Phagocyte	Multi-lobed nucleus in granular cytoplasm; engulfs bacteria	7000 (more during infections)	In red bone marrow		Actively seek and **engulf bacteria** – even squeeze through capillary walls to reach infected tissue. Often die loaded with killed bacteria. In boils this is seen as yellow 'pus'
Lymphocyte	Huge nucleus in little cytoplasm	2–3000 (more during infections)	In lymph nodes		React to proteins of invading organisms by making 'antibodies', which kill invaders and make their poisons (toxins) harmless (see unit 20.10)
Thrombocyte	Platelets are fragments of cells	¼ million	In red bone marrow		1 Stick to each other, forming a **temporary plug** in a cut blood vessel 2 Liberate an enzyme to help **clotting**; Ca^{2+} also needed: (a) fibrinogen → fibrin (soluble protein) (mesh of fibres) (b) fibrin traps blood cells which seal up the blood leak, prevent entry of harmful organisms, and dry to a protective *scab*, allowing healing of the wound beneath it

Platelet plug

Fibres

Cells trapped

Notes:

1 Haemoglobin combines 230 times more readily with carbon monoxide (CO) than O_2, forming a stable compound, **carboxy-haemoglobin** (Hb.CO), with it. Thus even at small concentrations in the air, CO (which is odourless) tends to be taken up into the blood, preventing O_2 from being carried. This can kill, e.g. someone tuning a car engine behind closed garage doors. See also *smoking* (unit 9.11).

2 Haemophiliacs ('bleeders') continue to bleed for a long time, even from minor wounds. They bruise easily and joints may be painful from bleeds. Whereas death was premature in the past, today haemophiliacs may live less dangerously by receiving the 'clotting factor VIII' which they lack (see unit 17.9).

8.3 Blood Smears

N.B.   The *slides* used should be rubbed clean with ethanol before use.
The skin must be sterilized with ethanol before *and* after extracting blood.
No lancet should be used by more than one person; its end must be sterile.

THIS WORK SHOULD ONLY BE DONE UNDER SUPERVISION OF A TEACHER

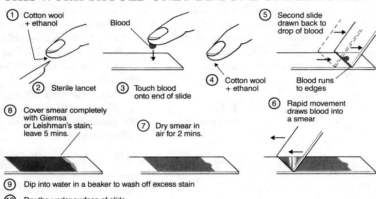

① Cotton wool + ethanol   Blood   ⑤ Second slide drawn back to drop of blood

② Sterile lancet   ③ Touch blood onto end of slide   ④ Cotton wool + ethanol   Blood runs to edges

⑧ Cover smear completely with Giemsa or Leishman's stain; leave 5 mins.   ⑦ Dry smear in air for 2 mins.   ⑥ Rapid movement draws blood into a smear

⑨ Dip into water in a beaker to wash off excess stain

⑩ Dry the *under*-surface of slide

⑪ Put a drop of glycerine on stained smear *where it is thinnest*

⑫ Under high power of the microscope, count the number of (a) red, (b) white blood cells in a single field of view at the *thinnest end* of the smear, where cells are evenly spread

⑬ Average the numbers for (a) and for (b), given by members of the class

⑭ Obtain a ratio of red to white blood cells

Fig. 8.2. Making a blood smear: staining and counting cells

Abnormal blood counts
Anaemia: red blood cell numbers down; exertion difficult.
Leukaemia: white blood cell numbers very markedly up; cancer of blood.

8.4 Blood Vessels and Blood Circulation

Table 8.2 Blood vessels and their functions

Arteries	Capillaries	Veins
Carry blood *away* from heart under *high* pressure	Carry blood from artery to vein, very slowly, giving maximum time for diffusion, through a hugh surface area.	Carry blood *towards* heart under *low* pressure
Carry *oxygenated* blood (except pulmonary artery) T.S. Elastic layer · Endothelium · Elastic and muscle layer	Endothelium only · Phagocyte emerging between cells of endothelium · 10 μm	Carry *deoxygenated* blood (except pulmonary vein) T.S. L.S. (a) free flow (b) back pressure Valve open Valve closed
(*a*) Heart refilling; elastic walls squeezing on blood to help it along (*b*) Heart pumping; 'pulse' felt as bore expands. Thick walls needed, but no valves Bore of arteries can be altered by nerve messages to muscle, e.g., more blood to legs and less to guts during exercise.	Tissue fluid leaking out to cells – blood pressure forcing it through See capillaries under the microscope in the tail of guppy fish or tadpole (head end in wet cotton wool)	No pulse: pressure is low at capillaries. Wall has 3 layers as in arteries but is thinner Blood returns partly by muscles of body squeezing veins – hence the need for non-return valves. Massage blood in an arm vein towards the fingers with the other thumb; valves show up as bumps (where they have closed)

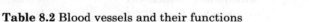

Portal veins have capillaries at either end, i.e. they carry blood from one organ to another (e.g. hepatic portal vein between small intestine and liver).

The heart and double circulation
The heart consists of two pumps fused together, each having an *auricle (= atrium) and a ventricle.*

The two pumps contract simultaneously according to a heart cycle (Fig. 8.5).

Right one pumps deoxygenated blood to the lungs for oxygenation.

Left one pumps oxygenated blood to the body, which deoxygenates it.

Thus blood passes twice through the heart before going to the body (see Fig. 8.3).

The resulting high blood pressure ensures:

(*a*) speedy supplies to the tissues;
(*b*) squeezing out of tissue fluid at capillaries.

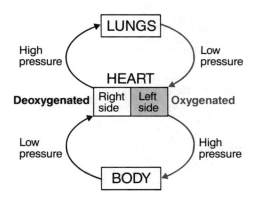

Fig. 8.3. Double circulation of blood through the heart of a mammal

8.5 The Heart

The heart lies between the two lungs inside the chest cavity.

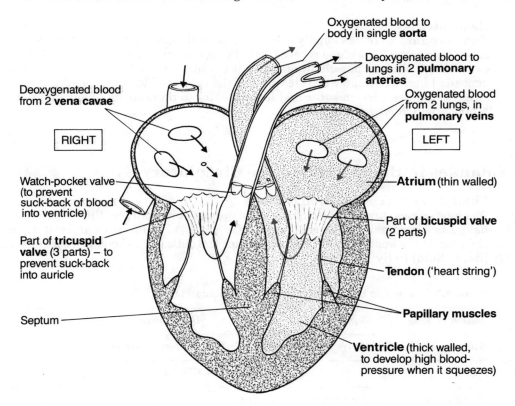

Fig. 8.4. The mammal heart in section: structure and function

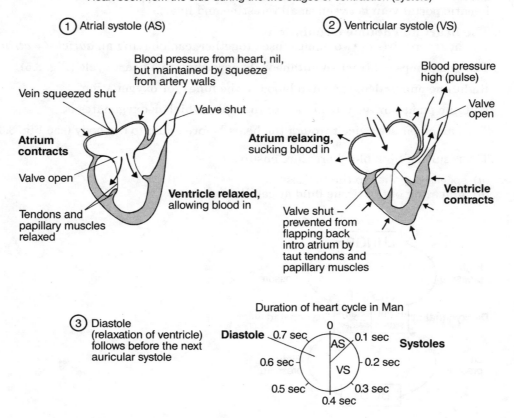

Fig. 8.5. The heart cycle of a mammal

Heart beat
Controlled automatically by a **pace-maker:** special tissue in the atrium wall. The rhythm speeds up when
(*a*) adrenaline (hormone) is secreted (see unit 12.10);
(*b*) nerve messages to the pace-maker arrive from the brain. A rise in blood CO_2 (from exercise) triggers this off.

Heart attack
Two **coronary arteries** supply heart muscle with blood. They exit just above the valve at the base of the aorta. This blood returns into the right atrium. Blockage of this mini-circulation may cause death of heart muscle by starving it of nutrients and oxygen. Fatty material (atheroma) deposited in the coronary arteries may be the cause of blockage. Smoking, excessive drinking, stress and a diet rich in saturated fats all seem to promote blockage and a high risk of heart attack – which may cause death.

8.6 Changes in Blood Around the Circulatory System

Changes in the composition of blood
As blood passes through the capillaries of organs, it is modified. Blood leaving endocrine glands has gained hormones, while that leaving the kidneys has lost urea and water. Thus *overall* blood composition is kept constant, ensuring that the cells of the body have a constant environment (tissue fluid) to live in.

Table 8.3 Changes in blood composition in the human body

Region of body	Blood gains	Blood loses
All tissues	CO_2; nitrogenous wastes	O_2; food; hormones
Lungs	O_2	CO_2; water
Small intestine	Food: water, salts, vitamins sugars, amino-acids	

Table 8.3 Changes in blood composition in the human body – contd.

Region of body	Blood gains	Blood loses
Liver	Controlled quantities of glucose and fats; urea	Glucose (for storage as glycogen); excess amino-acids; worn out erythrocytes
Kidneys		Urea; water; salts
Bones	New erythrocytes and phagocytes	Iron (for haemoglobin) Calcium and phosphate (for bone growth)
Skin	Vitamin D	Heat (by radiation and by evaporation of water in sweat). Salts and urea (in sweat)
Thoracic duct	Fats; lymphocytes; lymph	
Thyroid gland	Thyroxine	Iodine (to make thyroxine)

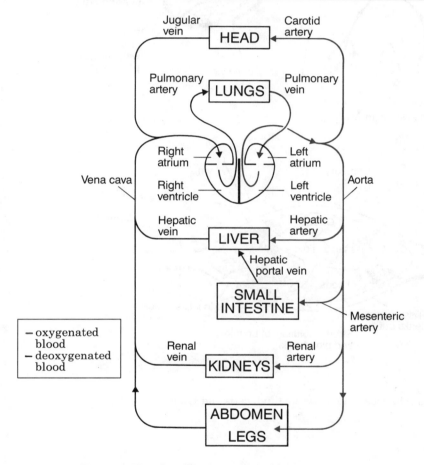

Fig. 8.6. The circulatory system of a mammal

8.7 Lymphatic System

A system of fine tubes ending blindly among the tissues, e.g. lacteals in villi of small intestine (see unit 6.9), which join up into ever larger tubes with non-return valves. Along their length are swellings (lymph nodes). The largest tube (thoracic duct) discharges into main vein of left arm.

Functions:

1 **Returns excess tissue fluid** to blood as lymph.

2 **Adds lymphocytes** to blood (for defence).

3 **Absorbs fats** (into lacteals of villi) to discharge them to blood.

4 **Filters out bacteria** from lymph by means of phagocytes stationary within lymph nodes (see Fig. 8.7).

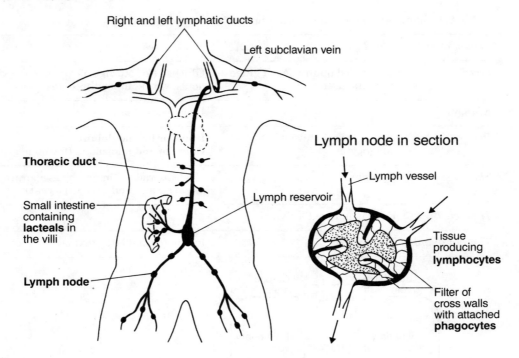

Fig. 8.7. The lymphatic system in Man

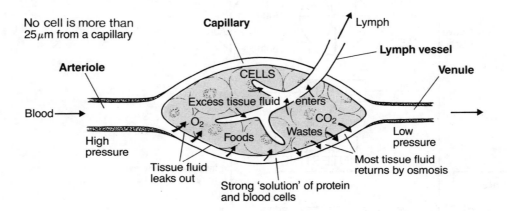

Fig. 8.8. The relationship between blood, tissue fluid, cells and lymph

9 RESPIRATION

9.1 Breathing, Gaseous Exchange and Cellular Respiration

Respiration is the sum of processes in organisms that leads to the release of energy from organic molecules, for use in vital functions. *All* organisms respire – plants as well as animals. Thereby they form ATP (see unit 9.14) – energy molecules that power the chemical reactions of metabolism. Depending on the kind of organism, up to *three processes* may be involved:

1 Breathing (= ventilation): *movements*, in animals, that bring a source of O_2 to a surface for gaseous exchange, e.g. chest movements of mammals bring air into lungs; throat movements in fish bring water (containing dissolved O_2) to gills.

2 Gaseous exchange: diffusion of O_2 into the organism and of CO_2 outwards. All gaseous exchange surfaces are moist, thin and have a large surface area.

(a) In *single-celled* organisms this exchange surface is the cell membrane (see Fig. 9.9).
(b) In *multicellular animals* specialized body parts, e.g. lungs, tracheoles or gills, provide the surface for gaseous exchange. Usually gases are transported rapidly by blood between these surfaces and a second extensive surface area where gaseous exchange occurs between the blood and cells (see Fig. 9.12).

Only insects pipe air directly to cells and do not use blood for this purpose (see Fig. 9.11).

(c) In *multicellular plants* a network of air spaces *between* cells allows for direct gaseous exchange between cells and the air. There is no blood system.

Thus gaseous *exchange* occurs only when organisms respire using oxygen.

3 Cellular respiration (= internal respiration): the chemical reactions occurring within cells that result in the release of energy to form ATP. These reactions can occur under two conditions:

(a) anaerobically – no oxygen needed (thus **1** and **2** above unnecessary);
(b) aerobically – oxygen needed (thus **2** above essential).

Note: since breathing and gaseous exchange are essentially *physical* processes occurring *outside* cells, they are often lumped together as **external respiration** to distinguish them from the *chemical* processes occurring *within* cells which are **internal respiration**.

Unfortunately the terms above are sometimes used loosely, e.g. since *Amoeba*, the earthworm and the flowering plant do not make *movements* to gain O_2, strictly speaking they do not *breathe* but they do respire.

9.2 Cellular Respiration (Aerobic and Anaerobic)

Glucose is the main substance respired (other foods can be turned into glucose). The results of respiration are different under anaerobic and aerobic conditions:

1 Aerobic
In plants and animals:

$$\text{glucose} + \text{oxygen} \xrightarrow[\text{and in mitochondria}]{\text{enzymes in cytoplasm}} \text{carbon dioxide} + \text{water} + \textbf{a lot of energy}$$
$$\text{(2890 kJ/mole)}$$
$$C_6H_{12}O_6 + 6O_2 \qquad\qquad 6CO_2 + 6H_2O$$

2 Anaerobic
(a) in plants:

$$\text{glucose} \xrightarrow[\text{matrix}]{\text{enzymes in cytoplasmic}} \text{ethanol} + \text{carbon dioxide} + \textbf{a little energy}$$
$$\text{(210 kJ/mole)}$$
$$C_6H_{12}O_6 \qquad\qquad 2C_2H_5OH + 2CO_2$$

(b) in animals:

$$\text{glucose} \xrightarrow[\text{matrix}]{\text{enzymes in cytoplasmic}} \text{lactic acid} + \textbf{a little energy}$$
$$C_6H_{12}O_6 \qquad\qquad 2C_3H_6O_3$$

Table 9.1. Comparison of the two stages in respiration.

	Anaerobic	*Aerobic*
Oxygen requirement	Nil	Essential
Useful energy from each glucose molecule respired	2 ATP	About 40 ATP
Chemical products	Organic, i.e. still energy-rich, e.g. lactic acid, ethanol	Inorganic: CO_2 and H_2O, i.e. no energy left
Takes place in	Cytoplasmic matrix (unit 1.2)	Mitochondria (unit 1.2)

Note: Aerobic and anaerobic respiration are *not* alternatives. Anaerobic reactions are the *first few stages* in a much longer set of reactions made possible under aerobic conditions (Fig. 9.1). Since aerobic respiration has the great advantage over anaerobic of providing about twenty times more energy, not surprisingly most organisms respire aerobically. Only certain bacteria cannot. However, some organisms are forced to respire anaerobically in their environment, e.g. tapeworms (see unit 21.15), or yeast in brewing operations. See Fig. 9.1 for a summary of respiration.

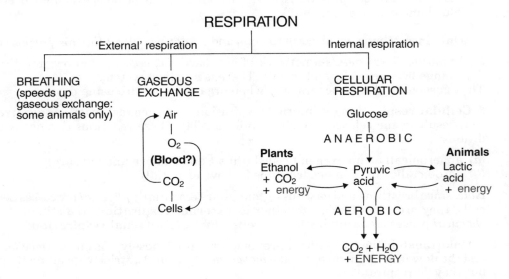

Note: Pyruvic acid (a 3-carbon compound) is common to all three respiratory pathways

Fig. 9.1 Respiration: breathing, gaseous exchange and cellular respiration

9.3 Anaerobic Respiration

Examples of anaerobic respiration in aerobic organisms

1 Man

(a) At rest, most of the pyruvic acid the cells produce is oxidized to CO_2 and H_2O. The blood contains very little lactic acid.

(b) During exercise, blood samples show that the lactic acid level rises at least ten-fold, indicating that despite increased breathing and heart rates, oxygen supply to tissues is inadequate. In this relatively anaerobic state Man is in **'oxygen debt'**.

(c) After exercise this debt is 'paid off' by continued rapid aerobic respiration. One fifth of the lactic acid is respired to CO_2 and H_2O. This provides energy to turn the other four-fifths of lactic acid back into glycogen (stored in liver and muscles).

How soon a person stops panting after exercise ('recovery time') is a measure of their **fitness**. During training, miles of extra capillaries grow, so increasing the oxygen supply to muscles. This increases muscle power, and reduces recovery time.

2 Yeast

(a) If aerated, the colony grows very rapidly in nourishing sugared water until all the glucose disappears as CO_2 and H_2O (no use to brewers!).

(b) Without air, in similar conditions, the colony grows more slowly, eventually killing itself in the ethanol it produces. This is the basis for *making wine and beer*. The ethanol can be distilled off (as in making *spirits*, e.g. whisky). This will burn, showing it is energy-rich.

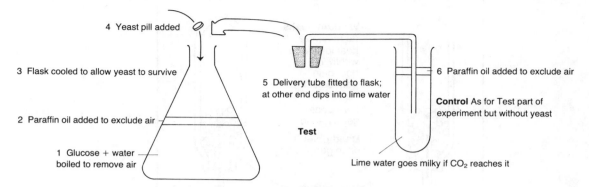

Fig. 9.2. Experiment to determine whether yeast will respire glucose anaerobically

3 Germinating peas
Half a batch of germinating peas is killed by boiling. Live and dead peas are washed in thymol solution to kill bacteria (which would produce CO_2). Both batches are put in boiling tubes in anaerobic conditions (see Fig. 9.3). Two days later the live peas have produced gas in an anaerobic environment.

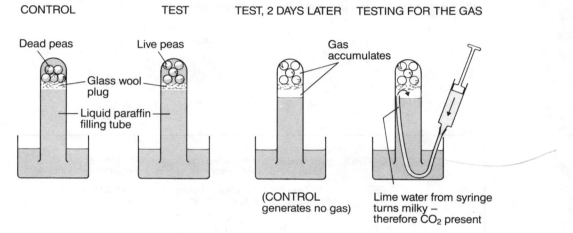

Fig. 9.3. Demonstration that germinating peas respire anaerobically

9.4 Aerobic Respiration

Three lines of evidence that organisms are respiring aerobically

1 **CO_2 evolved** (see Fig. 9.6).

2 **O_2 absorbed** (see Fig. 9.4).

3 **Heat** evolved. The energy in glucose is not totally converted into useful energy (ATP) during respiration. Some energy (around 60%) is wasted as heat (see Fig. 9.5). In the experiment shown in Fig. 9.5 both the dead peas (killed by boiling and then cooled for half an hour) and the live ones had been washed in thymol solution to exclude the possibility that bacterial respiration could be causing a rise in temperature. (An animal, e.g. locust, could be substituted for peas with similar, quicker results.)

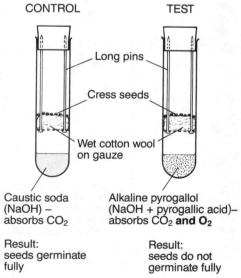

Fig. 9.4. Demonstration that seeds need oxygen to germinate

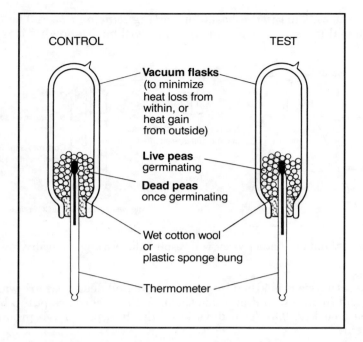

Fig. 9.5. Demonstration that germinating peas generate heat

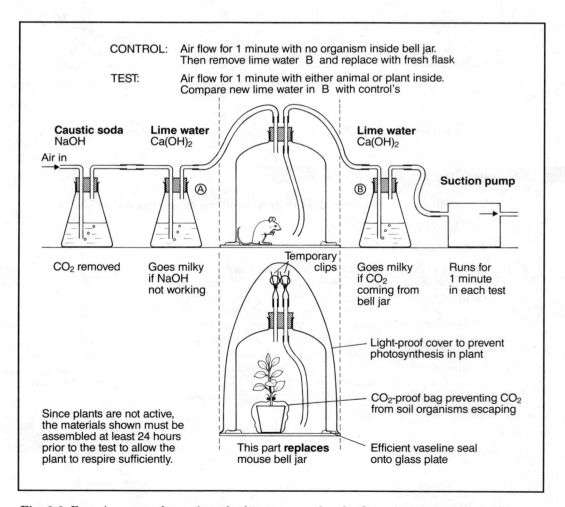

Fig. 9.6. Experiments to determine whether a mammal and a flowering plant produce CO_2

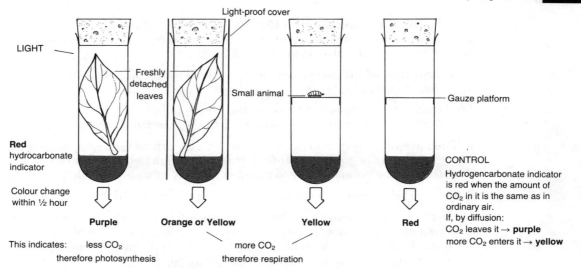

Fig. 9.7. Demonstration that animals and leaves respire, using an indicator method

9.5 Rate of Respiration

The rate of gaseous exchange can be used to find out the **rate of respiration** (Fig. 9.8).

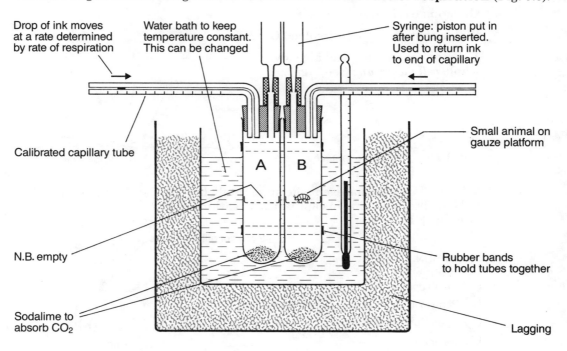

Fig. 9.8. Experiment to find out the rate of respiration, using a respirometer

Tube A will reach the temperature of the water in a few minutes. Insert its syringe plunger. Add an ink drop to the end of the capillary tube. No volume change (and therefore no movement of ink drop) is expected once the soda lime has absorbed the CO_2 in the air, since there is no organism respiring in it. Cooling or warming the surrounding water will, however, move the bubble. The distance the bubble moves in A must be subtracted from the distance moved in B.

Tube B is treated exactly as A is and at the same time. In B, however, the organisms take in O_2 and give out CO_2. Since the CO_2 is absorbed by soda lime, the volume of air in the tube becomes reduced. This causes more movement of the bubble, i.e. in addition to the movement, noted in B, owing to temperature change.

The respirometer can be used in water of different temperatures. The distance moved – in say 5 minutes – by the two ink drops is noted at each temperature.

9.6 Gaseous Exchange

All cells receive O_2 and lose CO_2 through **thin, moist membranes of sufficient surface area**. Multicellular organisms follow the same rules at their respiratory surfaces. They maintain a large surface area to volume ratio (see unit 22.6) to overcome the slow process of diffusion.

1 Cells must remain small if CO_2 and O_2 are to diffuse across the cytoplasm fast enough to maintain life. Cell division ensures this. Thus the volume of *Amoeba*, say 0.1 mm^3, is adequately served by its cell membrane area, say 2.5 mm^2.

2 A high rate of diffusion can be maintained by keeping a steep diffusion gradient (see unit 7.2). This happens by

(*a*) *breathing* (continually changing the air supply);
(*b*) *blood flow* (e.g. continually removing the O_2 absorbed into capillaries).

3 The rate of supply of O_2 and removal of CO_2 can be increased during exercise, e.g. in Man the breathing rate goes up four-fold, the volume inhaled per breath seven-fold, the heart rate doubles and the volume of blood pumped doubles or trebles (athletes can do better).

4 The surface area for gaseous exchange must remain high:

(*a*) *with air*, e.g. in Man about 700 million alveoli in his two lungs provide a total of about 80 m^2 (area of a badminton court)* to service his volume of about 80 dm^3;
(*b*) *with tissues*, e.g. in Man about 95 000 km of capillaries provide an area of about 700 m^2 of which about 200 m^2 are in use at any one time.

*Not tennis court size (261 m^2) as often quoted.

9.7 Organisms Respiring in Water and Air

Gaseous exchange in water

Water contains $< 1\%$ dissolved O_2.

1 Amoeba: gaseous exchange over whole *cell membrane* (see Fig. 9.9).

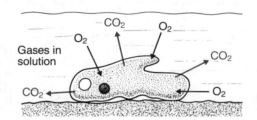

Fig. 9.9 Respiration in *Amoeba*

2 Bony fish: gaseous exchange at minutely branched *gill filaments* aided by blood containing red blood cells flowing in capillaries. Breathing requires use of mouth, pharynx and operculum (see Fig. 9.10).

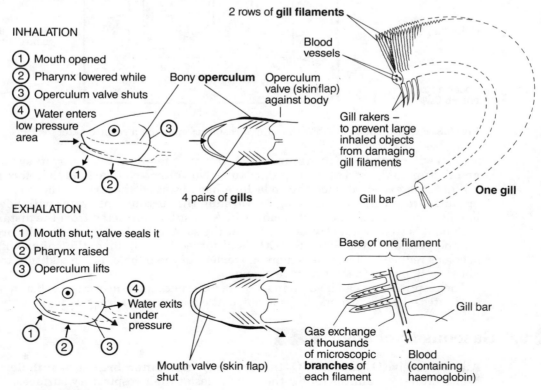

Fig. 9.10 Respiration in a bony fish

Gaseous exchange in air

Air contains almost 21% O_2.

1 Insect: gaseous exchange at *tracheoles*, thin tubes 1 μm in diameter supplying cells with air *direct* – blood not used for this. Much of the time O_2 and CO_2 just diffuse via *spiracles* and *tracheae* to tracheoles (see Fig. 9.11). Active or strong-flying insects, e.g. bees, and locusts, have air sacs. Abdominal *breathing movements* squash and unsquash these, assisting ventilation of tracheae. Locust group have a 'through system' for air (in through anterior spiracles, out through posterior).

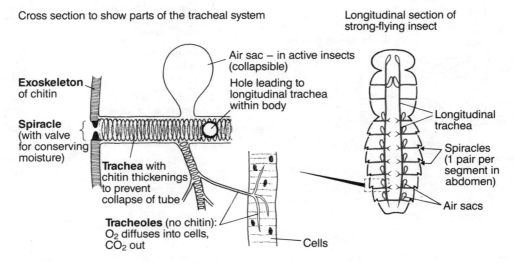

Cross section to show parts of the tracheal system

Longitudinal section of strong-flying insect

Exoskeleton of chitin

Spiracle (with valve for conserving moisture)

Air sac – in active insects (collapsible)

Hole leading to longitudinal trachea within body

Trachea with chitin thickenings to prevent collapse of tube

Tracheoles (no chitin): O_2 diffuses into cells, CO_2 out

Cells

Longitudinal trachea

Spiracles (1 pair per segment in abdomen)

Air sacs

Fig. 9.11 Gaseous exchange in an insect

9.8 Mammal Respiration

Gaseous exchange at millions of tiny air-sacs (*alveoli*). By diffusion, O_2 from air enters erythrocytes and CO_2 enters air from plasma. Flow of blood, in a network of capillaries around each alveolus, speeds up the exchange. The barrier to diffusion is slight – both the alveolus and the capillary walls are each only one cell thick (see Fig. 9.12).

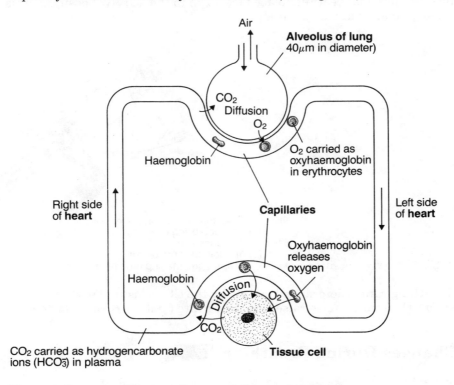

Air

Alveolus of lung 40μm in diameter)

CO_2 Diffusion

O_2

Haemoglobin

O_2 carried as oxyhaemoglobin in erythrocytes

Right side of **heart**

Left side of **heart**

Capillaries

Oxyhaemoglobin releases oxygen

Haemoglobin

Diffusion

O_2

CO_2

Tissue cell

CO_2 carried as hydrogencarbonate ions (HCO$_3^-$) in plasma

Fig. 9.12 Gaseous exchange at lungs and tissues of mammals

Note: When resting, breathing out occurs mainly because lung is elastic, collapsing if allowed to, thus deflating alveoli and bronchioles. Lungs may be made functionless by introducing air between pleural membranes, e.g. medically when treating tuberculosis (TB), or accidentally in a motor crash or a stabbing.

Table 9.2 Breathing movements.

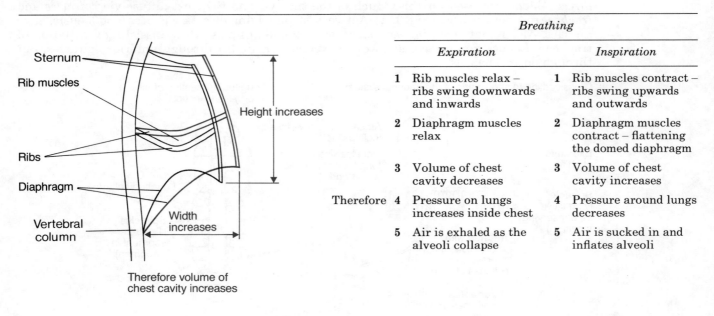

		Breathing	
		Expiration	*Inspiration*
	1	Rib muscles relax – ribs swing downwards and inwards	1 Rib muscles contract – ribs swing upwards and outwards
	2	Diaphragm muscles relax	2 Diaphragm muscles contract – flattening the domed diaphragm
	3	Volume of chest cavity decreases	3 Volume of chest cavity increases
Therefore	4	Pressure on lungs increases inside chest	4 Pressure around lungs decreases
	5	Air is exhaled as the alveoli collapse	5 Air is sucked in and inflates alveoli

Diagram labels: Sternum, Rib muscles, Ribs, Diaphragm, Vertebral column, Height increases, Width increases, Therefore volume of chest cavity increases

The kiss of life (mouth-to-mouth resuscitation)

A person suffering *asphyxia* (lack of oxygen, e.g. due to drowning or carbon monoxide poisoning) may need their breathing restored:

1 Force the victim's head back (to open the glottis) and pinch the nose shut.
2 Apply mouth to mouth and breathe forcibly into the victim's lungs.
3 Remove mouth for 4 seconds to allow the victim to breathe out.
4 Continue nos. 2 and 3 until breathing is restored.
5 Once the victim is breathing place face down in the 'survival position'; this prevents inhalation of any vomit that may be produced.

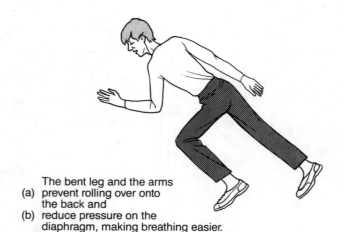

Fig. 9.13 The survival position

(a) The bent leg and the arms prevent rolling over onto the back and
(b) reduce pressure on the diaphragm, making breathing easier.

The breathing rate is determined mainly by the CO_2-sensitive part of the brain (see unit 12.7). A rise in CO_2 from exercise raises breathing rate (and heart rate) and depth of breathing.

9.9 Gas Changes During Breathing

Air breathed

Tidal air: about 0.5 dm^3 ($\frac{1}{2}$ litre) – quiet breathing at rest.
Vital capacity: about 3.5 dm^3 – volume inhaled or expelled in forced breathing.
Residual air: 1.5 dm^3 – air that cannot be expelled at all (remains in lungs).

Table 9.3 Approximate composition of air inhaled and exhaled (after removal of water vapour)

	Inhaled	*Exhaled*	*Approx. change*
Oxygen	21%	17%	20% decrease
Carbon dioxide	0.04%	4%	100-fold increase
Nitrogen	79%	79%	Nil

Air exhaled is also always saturated with water vapour (6%) – a variable loss of water from the body occurs, depending on how moist the inhaled air was.

9.10 The Respiratory Pathway

Air passes to alveoli via nostrils, nasal cavity, trachea, two bronchi with many branches, and millions of bronchioles (see Fig. 9.14). Dust, including bacteria, is 'filtered out' on sticky *mucus* in the nasal cavity as well as in the trachea. In both of them, *cilia* of lining cells pass the dirty mucus to the throat to be swallowed into the acid-bath in the stomach.

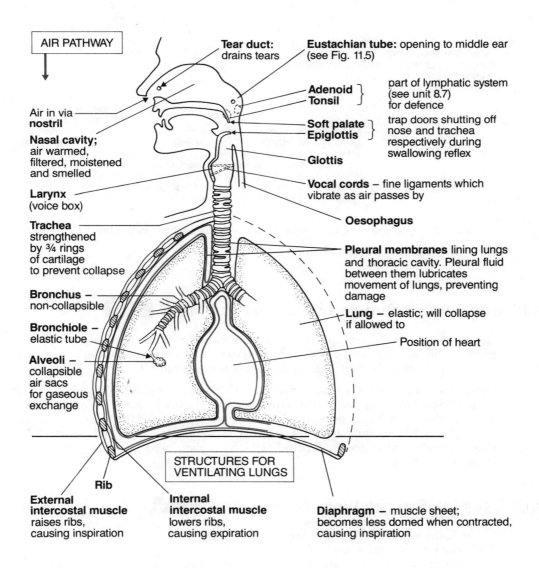

Fig. 9.14 Respiratory pathway in Man

9.11 Smoking or Health

Tobacco smoking has both short and long-term harmful effects (see Fig. 9.15).

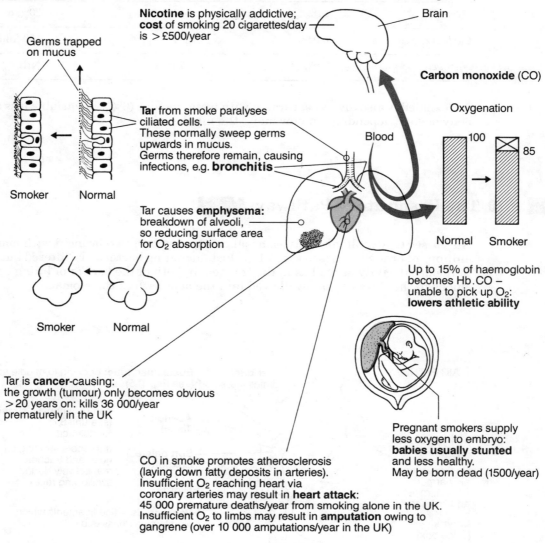

Nicotine is physically addictive; **cost** of smoking 20 cigarettes/day is > £500/year

Brain

Carbon monoxide (CO)

Oxygenation

Blood

Normal   Smoker

Tar from smoke paralyses ciliated cells.
These normally sweep germs upwards in mucus.
Germs therefore remain, causing infections, e.g. **bronchitis**

Germs trapped on mucus

Smoker   Normal

Tar causes **emphysema:** breakdown of alveoli, so reducing surface area for O_2 absorption

Smoker   Normal

Up to 15% of haemoglobin becomes Hb.CO – unable to pick up O_2: **lowers athletic ability**

Tar is **cancer**-causing: the growth (tumour) only becomes obvious > 20 years on: kills 36 000/year prematurely in the UK

Pregnant smokers supply less oxygen to embryo: **babies usually stunted** and less healthy. May be born dead (1500/year)

CO in smoke promotes atherosclerosis (laying down fatty deposits in arteries). Insufficient O_2 reaching heart via coronary arteries may result in **heart attack**: 45 000 premature deaths/year from smoking alone in the UK. Insufficient O_2 to limbs may result in **amputation** owing to gangrene (over 10 000 amputations/year in the UK)

Will-power and the use of nicotine chewing-gum are successful methods of breaking this addictive habit

Fig. 9.15 Effects of cigarette smoking

9.12 Gaseous Exchange in Flowering Plants

Angiosperms are flowering plants. Air diffuses through *stomata* (mostly on leaves; some on green stems) and *lenticels* on cork-covered roots and stems (see units 5.5 and 14.3, Fig. 14.5) to *air spaces* between cells, particularly of cortex and mesophyll.

Gaseous exchange: O_2 is absorbed and CO_2 released direct from cells to air spaces during **respiration** both day and night. However, *green cells* in sunlight absorb CO_2 and release O_2 during **photosynthesis** (see unit 5.5) at a rate far greater than the reverse process (owing to respiration). In dim light, e.g. dusk or dawn, rates of respiration and photosynthesis can be equal – the **compensation point**. No dead cells, e.g. xylem vessels (the majority in a big tree), respire or photosynthesize.

9.13 Uses for Energy from Respiration

1 **Mechanical** work, e.g. in contraction of muscles.
2 **Electro-chemical** work, e.g. in passing nerve impulses.
3 **Chemical** work, e.g. synthesizing large molecules, such as protein, from amino-acids during growth.

4 Heating, e.g. maintaining mammal body temperature.
5 Transporting, e.g. 'active transport' of materials across cell membranes (see unit 7.2).
Mnemonic: Make tea (MECH T)

9.14 ATP (Adenosine Tri-phosphate)

ATP is the 'energy molecule' of cells.
   When the end phosphate group is removed, leaving ADP (adenosine diphosphate), energy is released for use in any vital function, e.g. movement or growth. The phosphate may be added to ADP again, making ATP, during respiration in mitochondria. The energy for this comes from the energy in the sugar that is respired:

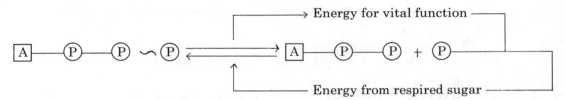

9.15 Measuring Energy Values of Foods

To measure energy in foods they must be dried and burned. During respiration, food is neither dry nor are there any flames. But the methods below are the best we can use (see Fig. 9.16).

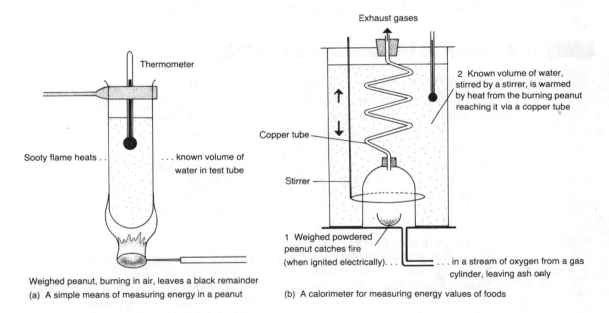

(a) A simple means of measuring energy in a peanut

(b) A calorimeter for measuring energy values of foods

Fig. 9.16 Measuring energy values of foods

   In (b) there is complete burning of the peanut, and better transfer of heat to water.

$$\text{The heat energy released (in kJ)} = \frac{\text{volume of water (cm}^3) \times \text{rise in temp (°C)} \times 4.2}{1000}$$

   The efficiency of transforming chemical energy of fuels and foods into mechanical energy (movement) can be judged from Table 9.4:

Table 9.4

	Efficiency of engines %			Efficiency %
	Steam	Petrol 4 stroke	Diesel	Respiration
Theoretical	30	58	65	40
Actual (approx.)	10	28	36	22

The wasted energy is lost as heat, in friction etc.

10 EXCRETION, TEMPERATURE REGULATION & HOMEOSTASIS

10.1 Wastes and Means of Excretion

Excretion is the removal of waste products of metabolism. Wastes are often toxic, particularly if they accumulate. Examples of excretion:

In **animals**

(i) CO_2 and water (from respiration);
(ii) ammonia, urea or uric acid (from protein metabolism – see unit 7.10).

In **green plants:**
(i) O_2 (from photosynthesis);
(ii) shedding leaves or bark (contain various wastes).

In **all organisms:**
   Heat energy (from metabolism – especially respiration, see unit 10.6). An important waste only in animals, when they move around. Loss of water unfortunately accompanies most forms of excretion.

Mammal excretory organs

1 **Lungs:** excrete CO_2; lose water vapour.
2 **Kidneys:** excrete urea; eliminate excess water and salts.
3 **Liver:** excretes bile pigments (see Fig. 6.13).
4 **Skin:** excretes some urea; loses water and salts (in sweat).

10.2 Mammal Urinary System

Two **kidneys** (see Fig. 10.1) at back of abdominal cavity:

(*a*) **excrete** waste **nitrogen** (from excess protein in diet) as urea;
(*b*) **eliminate** excess **salts** (e.g. NaCl in very salty food);
(*c*) **osmoregulate** to maintain **water** content of blood.

Blood pathway (see Fig. 10.1 Ⓑ)
Blood containing urea (made in the liver) passes into kidney from aorta via renal artery to about one million **glomeruli** (knots of capillaries); thence via further capillary network to renal vein and posterior vena cava. Blood is filtered at glomeruli.
   The filtrate produced is modified into urine as it passes through **nephrons** (filtration units). (See Fig. 10.2.)

Urine pathway (see Fig. 10.1 Ⓐ)
Urine formed by kidneys, is passed by peristalsis along two **ureters** to the bladder (storage); thence via **urethra** to outside (urination) (see Fig. 10.1 Ⓐ).
Urine in Man is a 2–4% solution of urea, some salts, yellow colouring (bile pigments accidentally absorbed in intestine), poisons, drugs and hormones (variously modified). Exact composition varies according to diet, activity and health. It is dilute if excess water is drunk; is concentrated after exercise. Normally about 1.5 dm^3 of urine is lost daily.

10.3 The Nephron

A nephron (see Fig. 10.2) is a kidney unit receiving tissue fluid and modifying it into urine. Tissue fluid (a filtrate of blood lacking cells and proteins) is forced out from the glomerulus (because of blood pressure) into the cavity of a **Bowman's capsule.** As filtrate passes along the tubules, all food and most other useful substances are reabsorbed from the tissue fluid, leaving urine.

10.4 Water Conservation

The **loop of Henle** makes the tissue fluid surrounding it salty (by secreting salt from the fluid it receives). This salty solution causes water to pass out of the **collecting duct** by osmosis (but *only* if the hormone ADH, which makes the duct's walls water-permeable, is secreted). ADH is only secreted if water needs to be conserved, e.g. owing to sweating. If *excess* water is drunk, it

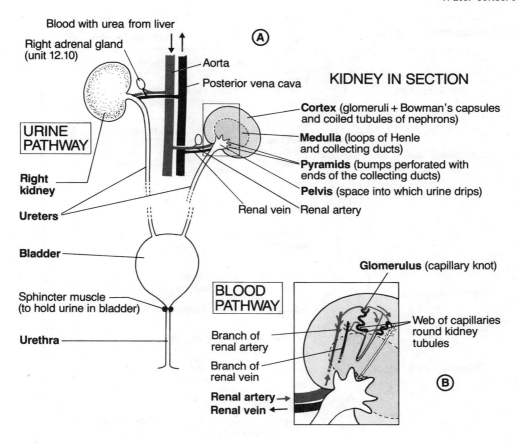

Blood with urea from liver

Right adrenal gland (unit 12.10)

Aorta

Posterior vena cava

URINE PATHWAY

KIDNEY IN SECTION

Cortex (glomeruli + Bowman's capsules and coiled tubules of nephrons)

Medulla (loops of Henle and collecting ducts)

Pyramids (bumps perforated with ends of the collecting ducts)

Pelvis (space into which urine drips)

Right kidney

Ureters

Renal vein Renal artery

Bladder

Glomerulus (capillary knot)

BLOOD PATHWAY

Web of capillaries round kidney tubules

Branch of renal artery

Branch of renal vein

Renal artery →

Renal vein ←

Sphincter muscle (to hold urine in bladder)

Urethra

Fig. 10.1 The mammal urinary system Ⓐ urine pathway from kidney Ⓑ blood pathway in the kidney

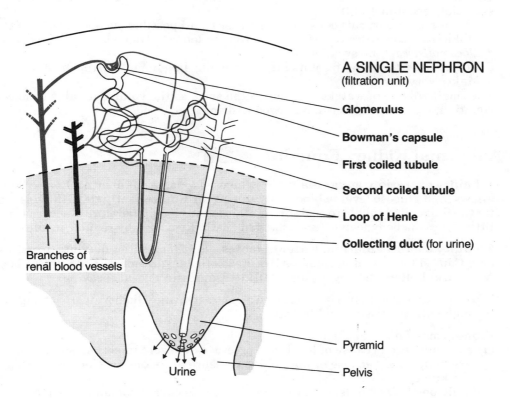

A SINGLE NEPHRON (filtration unit)

Glomerulus

Bowman's capsule

First coiled tubule

Second coiled tubule

Loop of Henle

Collecting duct (for urine)

Branches of renal blood vessels

Pyramid

Urine

Pelvis

Fig. 10.2 A single nephron of a kidney

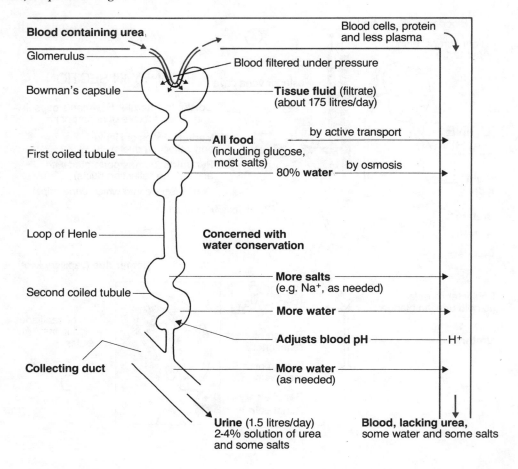

Fig. 10.3 How nephrons make urine in Man

passes out in the urine since it is not reabsorbed by the collecting duct owing to lack of ADH secretion (see unit 12.10).

The jerboa (desert rat) conserves its water by all possible means (compare rat in Fig. 7.12):
1 Makes *very concentrated urine* owing to long loops of Henle and high levels of ADH.
2 *Does not sweat* – no sweat glands.
3 *Evaporates little water* from its lungs – remains in a humid burrow by day.
4 Makes very *dry faeces.*
The small volume of water lost is much the same as the volume it gains from respiration (see equation, unit 9.2). Jerboas never need to drink.

10.5 Abnormal Kidney Function

1 **Faulty excretion: sugar diabetes** – glucose is passed out in urine. Lack of hormone **insulin** allows high glucose level in blood (see unit 12.10). Consequently tissue fluid is too glucose-rich for the first coiled tubule to reabsorb it all into the blood. Therefore glucose is drained, little by little, from the body; can cause coma and death. Remedied by regular insulin injections.

2 **Faulty osmoregulation: water diabetes** – large quantities of dilute urine, e.g. 20 litres per day. Caused by lack of hormone ADH. Leads to dehydration of body unless large volumes of water drunk. Remedied by regular ADH ('vasopressin') nasal spraying.

3 **Kidney disease: nephritis** – protein appears in urine. Glomeruli are letting plasma proteins through with the tissue fluid (filtrate).

Kidneys may fail
(*a*) suddenly, e.g. because of low blood pressure or severe infection (so killing the cells);
(*b*) gradually, e.g. because of high blood pressure or an obstruction preventing urine leaving the kidney.
   If only one kidney fails, the other healthy one is capable of doing the job of two.
There are three kinds of treatment for kidney failure:

1 Controlled Diet: Reduced intake of protein (less urea produced); less salt and water (less urine volume); and in particular less potassium-rich foods, e.g. oranges, chocolate, mushrooms (high K^+ can stop the heart).

If this fails to help and blood urea level rises (to five times the normal $0.3 \, g/dm^3$) the kidneys must be either assisted by dialysis or 'replaced' by healthy ones.

2 Dialysis by 'kidney machine': blood from an artery in the arm is passed through 10 m of dialysis ('Visking') tubing bathed in a special solution. This solution is similar to blood plasma but lacks protein and urea (see unit 8.2). The patient's urea and other wastes diffuse from the blood in the tubing into the bathing solution. The 'cleaned' blood returns to a vein in the arm. Fresh solution is used on every occasion.

It costs about £3000 per patient per year to give the necessary 12–18 hours of dialysis per week. The machines cost about £10 000.

A much simpler method of dialysis is 'CAPD' (body cavity dialysis). It allows the patient to remain active while dialysing (i.e. not attached to a machine.) A litre bag of glucose solution is drained into the body cavity. Wastes diffuse into the solution through the blood vessels of the gut. After some hours the solution is drained back into the bag and discarded.

3 Kidney transplant:
A healthy kidney (from a person only just dead or a living relative) is surgically inserted near the bladder. Certain precautions must be taken to avoid death of the transplanted kidney:
 (i) the *blood group* of the donor (giver) and recipient (receiver) of the kidney must be the same (see unit 17.3);
 (ii) if the *tissue type* of donor and recipient are also the same, success rate can be over 80%;
 (iii) the recipient's *antibody system* (see unit 20.10) must be suppressed by drugs for the rest of his or her life. This avoids rejection of the kidney but also risks serious illness from other, ordinary, infections. So antibiotics are often also given.

10.6 Body Temperature in Organisms

Skin and temperature control
1 The *body generates heat* by its metabolism (60% of the energy from respiration is wasted as heat), e.g. blood leaving contracting muscles or the liver is warmer than when it entered them.
2 At the *skin*, blood either *loses* this heat to cooler surroundings or *gains* even more if the surroundings are warmer.
3 *Gain or loss of heat* can happen in four main ways (see Fig. 10.4):
(*a*) radiation (important in air) – Man, at rest in shade, loses most this way.
(*b*) conduction (important in water) – e.g. elephants bathing.
(*c*) convection – air circulation; speeds up (*a*) and (*b*).
(*d*) evaporation (heat *loss* only – heat transfers to water which gains enough energy to vaporize. This happens during breathing, panting and sweating in animals; and during transpiration in plants).
Some heat is also lost in *urine* and *faeces*.

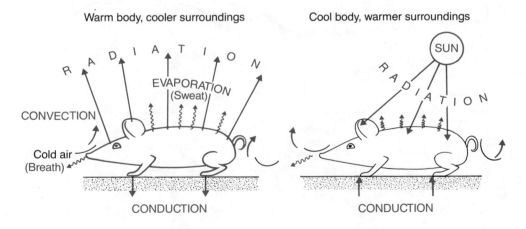

Fig. 10.4 Heat gain and loss by a mammal

4 Thus *most animals* (and all plants) have body **temperatures that fluctuate** with that of their environment. These animals are called poikilotherms or **ectotherms** or 'cold-blooded'.
Birds and mammals have body **temperatures that remain constant** despite the fluctuating environmental temperature. They are called homoiotherms or **endotherms** or 'warm-blooded'.

Table 10.1 Comparison of ectotherms and endotherms

	Ectotherms	*Endotherms*
In cold conditions	Become sluggish as they cool because their enzymes work more slowly. To avoid death by freezing may need to **hibernate**	Can remain **active** even in polar regions because their enzymes are kept working at their best (optimum) temperature. Only small mammals, e.g. dormice, marmots need to hibernate
In hot conditions	Active, but may need to **aestivate** to avoid over-heating, e.g. earthworm curls up into an inactive ball deep in soil	**Active** – cooling measures work; none aestivate. Some avoid heat by being active at night (cooler)
Main disadvantage	**Fall easy prey to endotherms** when not fully active – particularly when hibernating or aestivating	**Require a lot more food** to keep up their temperature

10.7 Mammal Temperature Control

Mammals have a *thermostat* in the fore-brain (see unit 12.7) which monitors blood temperature. Its information causes changes in:

1 Behaviour: e.g. seeking shade or getting wet if it is hot; seeking shelter and huddling into a ball if it is cold (see Fig. 10.5).

2 Skin (physical control):
(*a*) **hair:** traps air – a good insulator. Amount of insulation can be varied by raising or lowering hair using erector muscles. Thicker 'coats' in winter – moulted in summer.
(*b*) **fat:** also a good insulator. Whales (in very cold water) have thick 'blubber' but camels have no fat except in hump. Mammals prepare for winter cold by laying down more fat.
(*c*) **capillaries and shunts:** skin 'flushes' with blood flowing through surface capillaries (*vaso-dilation*) which radiate heat when mammal is hot. Skin goes pale if cold since blood is diverted from surface capillaries (*vaso-constriction*) often by going through a 'shunt' deeper down.
(*d*) **sweat glands:** secrete sweat (salty water containing some urea). Water evaporates, removing excess heat.

3 Metabolic control:
(*a*) **shivering:** involuntary contractions of muscles generate heat.
(*b*) **liver:** metabolizes faster owing to increased thyroxine secretion (see unit 6.11).

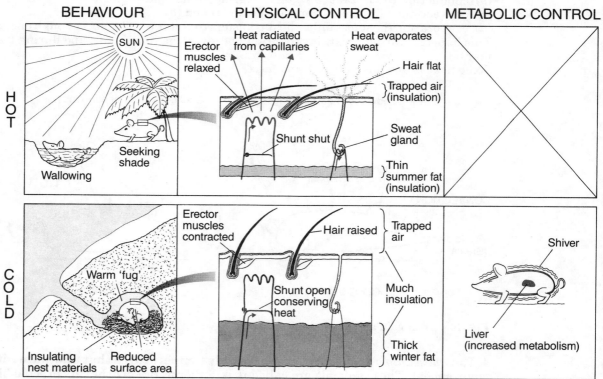

Fig. 10.5 Three ways of maintaining constant body temperature

4 **Shape:** a large surface area to volume ratio (see unit 22.6) assists heat exchange, e.g. body stretched out to lose heat. The opposite, e.g. body curled up, conserves heat. For the same reasons, large ears of desert foxes and hares radiate heat well; small ears of arctic foxes and hares radiate less (see p.119).

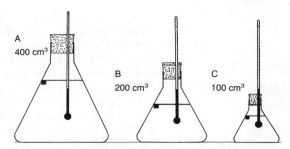

Fig. 10.6 Experiment to discover the importance of surface area to volume ratio in the cooling of water

Flasks A, B and C are filled with water at, say, 60°C. Their temperature is read every 2 minutes and recorded. The surface area is measured by shaping and cutting graph paper to fit the outside and counting squares.

Results show that C cools fastest and A slowest.

This may help explain why small mammals, e.g. shrews, need to eat so much for their size (respire a large part of their food to maintain temperature).

10.8 Temperature Control in Other Organisms

1 **Ectotherms:** rely on behaviour to keep a constant temperature – move to warm or cold places as the situation demands. Do not use skin or metabolic means as mammals do. However, wood-ant nests (28°C) and honey-bee hives (35°C) *are* maintained at the temperatures indicated.
2 **Angiosperms:** cannot move; they perennate if temperature becomes impossible (see unit 14.3). If hot they *transpire* more (water evaporation) or *wilt* reducing area of leaves gaining heat from sun.

10.9 Homeostasis

Homeostasis is the maintenance of a constant environment immediately around cells. For unicellular organisms this is the water they inhabit and their only means of homeostasis is to move (if they can) to a suitable area. The immediate environment of cells in a multicellular animal is the tissue fluid. In mammals the composition of this is kept very constant by a variety of organs, each of which controls particular factors in the blood (the source of tissue fluid).

Table 10.2 Organs concerned with homeostasis in Man

Organs concerned	Factors controlled in blood	Healthy blood levels in man
Liver and islet tissue of pancreas (unit 6.11)	Glucose	1 g/dm³
Skin, liver (Fig. 10.7)	Temperature	36.8°C (under tongue)
Kidneys (unit 10.3)	Osmoregulation (water) pH (acidity/alkalinity) Urea (nitrogen waste)	90% pH 7.4 0.3 g/dm³
Lungs (unit 9.8)	Carbon dioxide (carbon waste) Oxygen	550 cm³/dm³ (at rest, deoxygenated) 193 cm³/dm³ (at rest, oxygenated)

Note: blood does of course vary in composition according to where it is in the body (see unit 8.6), but overall the levels of factors affecting the vital functions of cells are kept within narrow limits.

10.10 Skin Functions

1 **Sensory:** sensitive nerve endings give warning of harm – pain, touch, heat or cold.
2 **Protection:** skin acts as barrier between the internal environment of cells (tissue fluid) and the external environment (anything from climate, air or water to bacteria or predators).
  Skin *resists:*
  (*a*) **puncture** – (from slashes, blows or friction) by being tough and hair-padded;
  (*b*) **desiccation** – (drying of body) by the waterproof protein keratin, aided by oils;
  (*c*) **entry of pathogens** – (viruses, bacteria, etc.);
  (*d*) **damage from ultra-violet light** – ('sunburn'; skin cancer) by suntanning, i.e. producing more pigment when in sunshine.

Skin *assists* predators and prey by providing:

(*e*) **weapons** from modified skin – (claws, hooves) for attacking or defending;

(*f*) **camouflage** – by special distribution of pigment in three ways:

(i) *blending:* similar colour to background, e.g. khaki colour of lion.

(ii) *countershading:* pale belly is darkened by shadow; dark back is made paler by sun.
Therefore from the side the animal looks 'flat'; difficult to see, e.g. deer.

(iii) *disruptive:* regular outline broken up by stripes or blotches to blend with light and shade among vegetation, e.g. leopard.

3 Synthesis: certain oils in the skin are changed to *vitamin D* (see unit 4.5) when subjected to ultra-violet light.

4 Excretion: some *urea* is lost in sweat.

5 Temperature control is dealt with in unit 10.7.

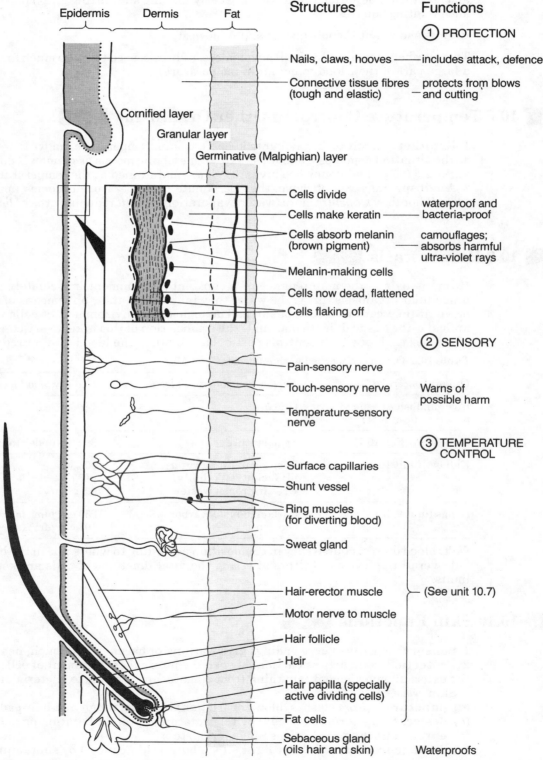

Fig. 10.7 The structure and functions of mammal skin

11 SENSITIVITY

11.1 Sensitivity in Plants and Animals

Organisms must be aware of their surroundings and respond to them, where necessary, to keep alive. Plants must seek light; animals, food. Organisms respond to various **stimuli** (detectable changes in the environment). Plants respond to light, gravity, touch (see unit 12.14); and animals also respond to these and to temperature, chemicals in air (smells) or water (tastes) and sound. Plants use much simpler means than animals to detect and respond to stimuli (see unit 12.1).

Table 11.1 Comparison of sensitivity and response in animals and plants

Multicellular animals	*Multicellular plants*
1 **Special sense cells** or organs (which usually do nothing else – e.g. eyes which only see)	No *special* sense organs, e.g. shoot tips sense light
2 **Nerves** relay messages from sensory areas	No nerves
3 **Brain** (present in most) 'computes a decision', sent to muscles	No brain
4 **Muscles** which can move the whole body towards or away from the stimulus	No muscles: cannot move the whole body

11.2 Mammal Sense Organs

Sense organs pick up stimuli – they *sense*. But sensations are interpreted (*perceived*) by the brain, e.g. eyes may work perfectly but if the optic nerve or the visual centre of the brain is damaged, the person is blind.

Sense organs sense stimuli in both the external and internal environments:

External

(*a*) *Skin* – touch, heat, cold, pressure (extremes of which can cause pain) (see unit 10.10).
(*b*) *Nose* – air-borne chemicals (smells – including the 'taste' of food).
(*c*) *Tongue* – chemicals causing perception of bitter, sweet, salt and sour tastes.
(*d*) *Ear* – sound (high frequency pressure changes); changes of body position; gravity sense.
(*e*) *Eye* – light (as light or dark, colour, and the form of objects).

Internal (often concerned with homeostasis, see unit 12.9). Examples:

(*a*) *Thermostat* in hypothalamus of brain (see unit 10.7).
(*b*) *Breathing centre* (CO_2-sensitive) in medulla oblongata of brain (see unit 9.8).
(*c*) *Spindle organs* sense tension in muscles. These assist muscle coordination and supporting of loads (see Fig. 11.1a and b).
(*d*) *Tendon organs* sense tension in tendons. These prevent overloading of muscles and tendons (see Fig. 11.1a and b).

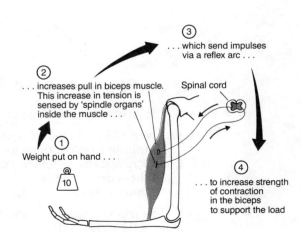

Fig. 11.1 The role of stretch receptors in the arm: 'feedback' to muscles (a) Supporting a load

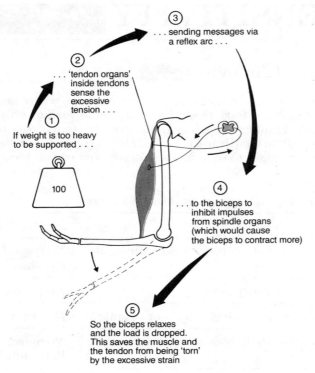

③ ... sending messages via a reflex arc ...

② ... 'tendon organs' inside tendons sense the excessive tension ...

① If weight is too heavy to be supported ...

100

④ ... to the biceps to inhibit impulses from spindle organs (which would cause the biceps to contract more)

⑤ So the biceps relaxes and the load is dropped. This saves the muscle and the tendon from being 'torn' by the excessive strain

Fig. 11.1 The role of stretch receptors in the arm: 'feedback' to muscle. (b) Dropping a load

11.3 The Eye

1 The outer covering of the eyeball is the tough, white **sclera**. It joins with the transparent **cornea** in front. Both are kept in shape by pressure from tissue fluid secreted from **ciliary body** capillaries (see Fig. 11.2).

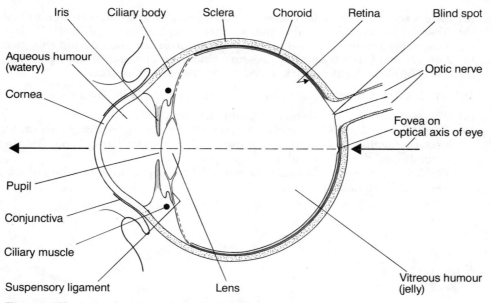

Iris   Ciliary body   Sclera   Choroid   Retina   Blind spot

Aqueous humour (watery)

Optic nerve

Cornea

Fovea on optical axis of eye

Pupil

Conjunctiva

Ciliary muscle

Suspensory ligament   Lens   Vitreous humour (jelly)

Fig. 11.2 Horizontal section through the left eye of man

2 It is **protected** within a bony socket (*orbit*) and by **three reflexes:**

(a) *weep reflex: dust and irritants* sensed by the **conjunctiva** cause an increase in tears and blinking to wash them away;

(b) *iris reflex: strong light* on the *retina* causes a narrowing of the pupil to prevent damage to the light-sensitive cells;

(c) *blinking reflex:* seen *objects* which may hit the head cause the eyelids to close.

3 The **iris** is a muscular sheet bordering a hole, the **pupil**. Its size is controlled by two sets of muscle (see Fig. 11.3).

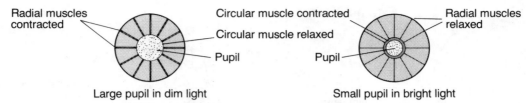

Fig. 11.3 View of the iris from within the eye

4 The **retina** contains nerve cells linked with two kinds of **light-sensitive** cell:
(*a*) *rods:* sense in black-and-white, in dim light; most are outside fovea.
(*b*) *cones:* sense in colour, in brighter light; very high numbers at fovea.
The *fovea* (called the 'yellow spot' in Man) enables objects to be seen in colour and in great detail. Image is only in complete focus at this spot. Is virtually blind in very dim light.
The *blind spot* has only nerve cells (gathering into the *optic nerve*): no vision.

5 The **choroid:**
(*a*) Black pigment cells prevent internal reflection of light.
(*b*) Capillaries supply the retina with tissue fluid.

6 Focusing
Cornea: responsible for most (at least 70%) of the converging of light rays.
Lens: makes the final adjustment, i.e. *accommodates.*
  Far-focusing: lens pulled thin by strain on suspensory ligaments exerted by sclera under pressure from tissue fluid.
  Near-focusing: lens collapses fat owing to its elasticity, when strain on suspensory ligaments is reduced by contraction of ciliary muscle.

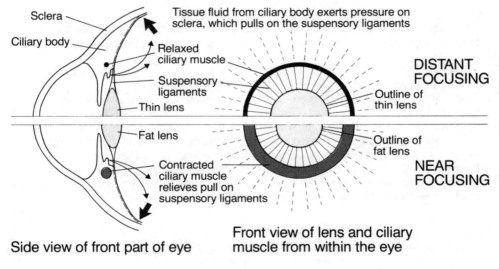

Fig. 11.4 Focusing in the eye

7 Binocular vision: the two different views obtained by human eyes overlap. The brain 'computes' how far away objects are.
  Six muscles, attached to the sclera of each eye, swivel the eyes to look at the same object.

11.4 The Ear

Table 11.2 Functions and parts of the ear

Functions	*Parts*
Hearing	Outer: ear *pinna* and *canal* for sound-gathering
	Middle: *ear-drum* vibrates; *3 ossicles* transmit vibrations; together they amplify the sound at *oval window*
	Inner: (*a*) *cochlea* (a 3-part spiral tube filled with liquid) receives the amplified sound waves which stimulate *hair cells* in the middle tube
Detecting change in position	Inner: (*b*) *3 semicircular canals* (set at right angles to each other) whose liquid moves inside *ampullae* (swellings), so stimulating *hair cells* there

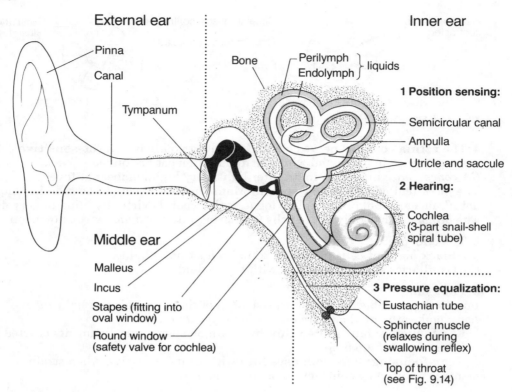

Fig. 11.5 Structure and function of the three parts of the ear and Eustachian tube

Messages from both sets of hair cells go via the *auditory nerve* to the brain: the *cerebrum* interprets sounds received and the cerebellum contributes to sense of balance (see Fig. 12.6).

Hearing

Man can hear sound *frequencies* of 20–20 000 Hertz.

Volume can be amplified up to 22 times: pressure of sound waves reaching the tympanum is transmitted by the ossicles to an area 22 times smaller – the base of stapes. Muscles acting on ossicles can also diminish their movements during loud noise, preventing ear damage.

The stapes vibrates because the tympanum does; thus air-borne sound waves are changed into liquid-borne sound waves in perilymph. These waves cause movement of a membrane on which hair-cells sit. These hairs, embedded in a jelly shelf, are pulled or squashed as the cells rise or fall on the membrane, sending impulses along the auditory nerve. This sound-sensing part of the cochlea is called the organ of Corti (see Fig. 11.6).

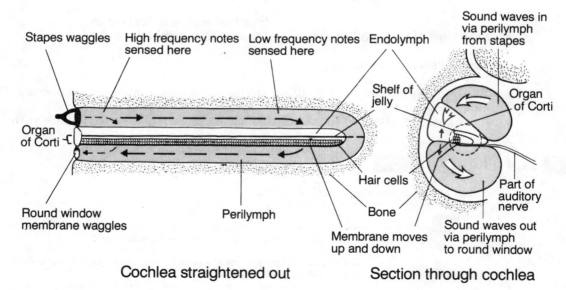

Fig. 11.6 How sounds are sensed in the cochlea

Detecting change in position

(a) Inside each **ampulla** are hair cells capped by a cone of jelly (cupula). The cupula can move like a swing-door. It swings only when its semicircular canal and the head are moving in the same plane. Although the canal moves, the fluid does not (owing to inertia). So the cupula swings, bending the hairs, which send impulses to the brain (see Fig. 11.7). According to which of the 6 semicircular canals (in 2 ears) are stimulated, the brain can 'compute' how to keep balance. Information from the eyes assists in this.

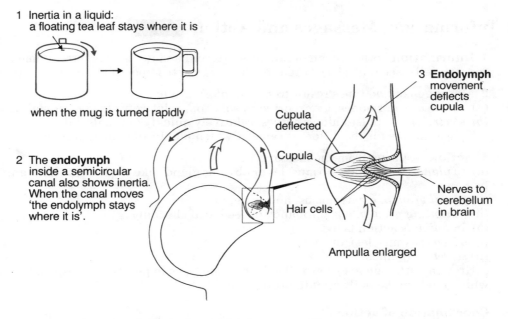

1 Inertia in a liquid:
  a floating tea leaf stays where it is

when the mug is turned rapidly

2 The **endolymph** inside a semicircular canal also shows inertia. When the canal moves 'the endolymph stays where it is'.

Cupula deflected

Cupula

Hair cells

3 **Endolymph** movement deflects cupula

Nerves to cerebellum in brain

Ampulla enlarged

Fig. 11.7 Detection of change in position by an ampulla

(*b*) The **utricle** and **saccule** also each contain hair cells with a cupula. The cupulae are weighted with chalk grains. As the head tilts forwards or backwards the weight of the cupula pulls more on the hair cells of the utricle. These send impulses to the brain via nerves. The saccule deals with sideways tilts of the head. This way we know whether we are upright or at a tilt (see Fig. 11.8).

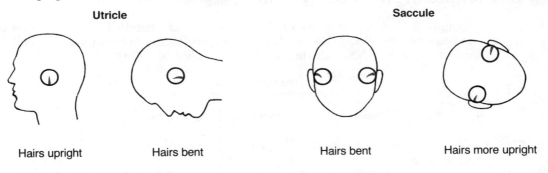

Utricle

Hairs upright

Hairs bent

Saccule

Hairs bent

Hairs more upright

Fig. 11.8 The sense of position at rest

12 COORDINATION AND RESPONSE

12.1 Information, Messages and Action

1 Information both from an organism's external and its internal environments is received by sensory cells (see unit 11.1). Often this has to be acted upon if the organism is to remain alive.

2 Messages of two types result from the information:
(*a*) *chemical* – hormones, transported in solution, relatively slowly (animals and plants);
(*b*) *electrical* – impulses along nerves, relatively quickly (animals only).
This accounts for the different rates at which plants and animals react.

3 Action resulting from the message:
(*a*) in *plants* (which have no muscles or obvious glands like the liver, as have animals) is usually by:
(i) *special growth*, e.g. tropisms, flowering, or
(ii) *inhibiting growth*, e.g. dormancy of seed, leaf shedding;
(*b*) in *animals* action is by:
(i) *movement* (muscles), or
(ii) *secretion* (glands).

Growth, although also controlled by hormones, as in plants, is a response only to the rate at which food can be built up into protoplasm.

Coordination of actions
Each response to a stimulus, unless coordinated with others, would lead to chaos. Thus feeding on bread includes muscle coordination to get the bread into the mouth (and not the ear) and to cause chewing, swallowing and peristalsis, as well as coordination of secretion of saliva, mucus and pancreatic juice (at the right times).

12.2 Mammal Nervous System

Composed of *neurones* (nerve cells). Neurones are bundled up into *nerves* in the *peripheral nervous system (PNS)*. Nerves link sensory cells and action (effector) cells with the *central nervous system (CNS)* – the brain and spinal cord (see Figs. 12.1, 12.2).

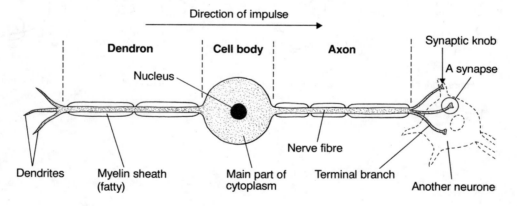

Fig. 12.1 A generalized neurone

As far as function is concerned there are *four types of neurone:*

1 Sensory: may connect with sensory cells, e.g. in retina of eye, or have sensory ends themselves, e.g. touch receptors in skin. Have long dendrons, short axons; carry 'messages' about the environment *to* the CNS.

2 Relay: always act as links between neurones, e.g. sensory neurones and either motor neurones or pyramidal neurones. Thus allow a large number of cross-connections, as in a switch-board.

3 Motor: usually link with relay neurones and with muscle or gland cells, to which they carry 'messages' *from* the CNS, calling for action; have short dendrons, long axons.

4 Pyramidal: connect with relay neurones and other pyramidal neurones which have a vast network of cell branches (up to 50 000), each a possible inter-connection. This allows the 'computer' function of the brain (see Fig. 12.4).

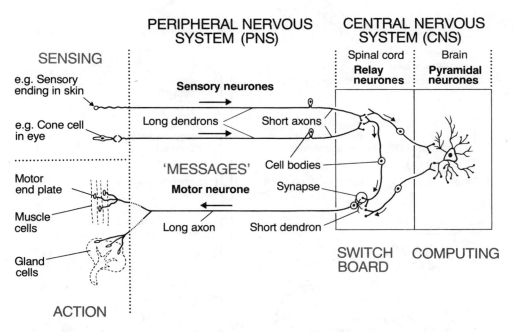

Fig. 12.2 Four different kinds of neurone and their functions in the body

12.3 Nervous Impulses

An impulse ('message') is produced by the flow of Na^+ ions into a neurone and K^+ out. It passes along the neurone at up to 120 m/s.

Impulses *cause secretion* of a chemical substance at a *synaptic knob* which, for less than one millisecond, 'connects' two neurones electrically, allowing the impulse to pass on. The chemical is destroyed and re-created after each impulse (see Fig. 12.3). Thus neurones are not physically connected to each other (as in an electrical circuit) and each neurone generates its own electricity (there is no central battery). Each synapse is effectively a connecting switch.

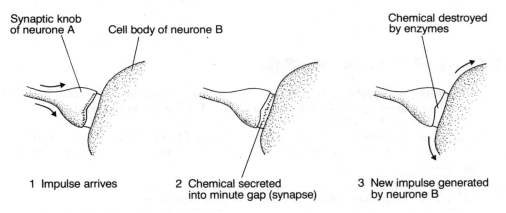

Fig. 12.3 Passing a 'message' at a synapse

12.4 Reflex Action

Reflex action: an automatic rapid, *unlearned* response to a stimulus which helps the animal survive. It is a reaction to sensory information of an *urgent* nature (e.g. withdrawing hand from flame; righting oneself when overbalancing; swallowing) which could mean the difference between survival and death (see also unit 11.3 – eye).

A maximum of *five* kinds of cell (*reflex arc*) take part in a reflex action (see Fig. 12.4, numbers ① – ⑤).

A knee-jerk reflex arc has no relay neurone.

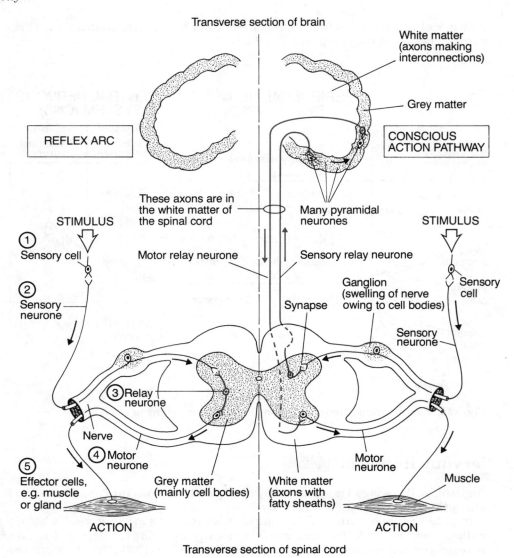

Fig. 12.4 Comparison of reflex and conscious action pathways

12.5 Learned Behaviour

1 Conditioned reflex action: a *learned* reflex, i.e. the brain is involved. During the training period an inappropriate stimulus is substituted for the appropriate one, as Pavlov discovered with dogs (see Table 12.1).

Conditioned reflexes can be 'unlearned' too, if the reaction is not rewarded. Many skills, e.g. feeding oneself, writing, riding a bicycle, are conditioned reflexes learned by hard practice (training).

Table 12.1 Pavlov's experiment – conditioned reflex in dogs

	Stimulus	*Reaction*
Reflex action	Smell of food	Saliva flows
Training period	Bell rung when food given	Saliva flows
Conditioned reflex	Bell rung (inappropriate)	Saliva still flows

2 Conscious action: sensory information goes to the brain before action is taken (see Fig. 12.4). All the little delays in transmission of impulses at thousands of synapses in the brain add up to make reaction time slower than in reflex actions.

Animals with small brains rely mostly on automatic reactions (instinct). Those with larger brains have more scope for working out solutions (intelligence).

12.6 Instinctive Behaviour

Instinctive behaviour: a series of unlearned actions, the completion of one being the signal for starting the next. Often a highly complex 'behaviour pattern', any disruption of which leads to total failure and a recommencement of the sequence of actions, e.g. provision of food for a hunting-wasp's larvae (see Fig. 12.5).

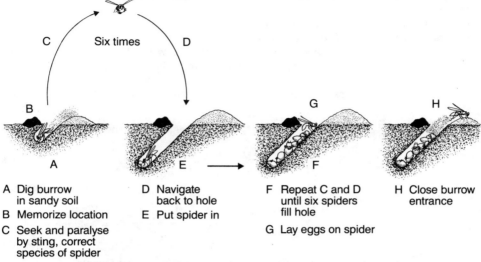

A Dig burrow in sandy soil
B Memorize location
C Seek and paralyse by sting, correct species of spider

D Navigate back to hole
E Put spider in

F Repeat C and D until six spiders fill hole
G Lay eggs on spider

H Close burrow entrance

If at stage E spiders are removed by forceps, wasp continues to bring more spiders, eventually giving up and starting at A again, elsewhere

Fig. 12.5 Instinctive behaviour of a hunting wasp in providing food for its larvae

12.7 The Brain

Expanded front part of nerve cord; but grey matter is outside the white. In primitive vertebrates, the brain has three main parts; fore-, mid- and hind-.

In most mammals the same three parts are easily seen.

In man, fore-part (cerebrum) is so vast that it covers mid-brain and part of hind-brain too (see Fig. 12.6).

1 Fore-brain
(a) **olfactory lobes** (in front) – sense of smell.
(b) **cerebrum** (upper part) – centre for memory, aesthetic and moral sense, hearing, vision, speech and muscular action other than in the guts and blood vessels.
(c) **hypothalamus** (lower part) – receptors for control of internal environment (homeostasis), e.g. temperature, water content of blood. An outgrowth of it is the **pituitary** (the 'master' endocrine gland, see unit 12.10).

2 Mid-brain
Optic-lobes (upper part) – simple auditory and visual (pupil) reflexes.

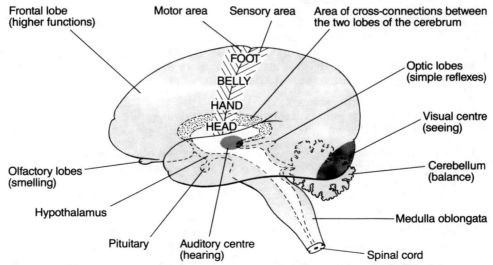

Fig. 12.6 Functional areas of the human brain. (Fore-brain shown overlying mid-brain and hind-brain in section)

3 Hind-brain
(a) **cerebellum** (large upper outgrowth) – balance, coordination of muscle action.
(b) **medulla oblongata** (brain-stem, merging with spinal cord behind) – control of many vital 'automatic' actions, e.g. breathing, heart rate, constriction of arteries to direct blood to specific regions of the body etc.

Summary of brain functions: receives all sensory information and 'processes' it (see Fig. 12.7) either:
(a) immediately – reflex action (as in spinal cord), or
(b) more slowly – *storing* it as 'memory'
    – using past memory to compare with the new, and *calculating*
    – *coordinating* memories from other brain centres
    – reaching a *'decision'*
    – passing out *'orders'* via neurones; and hormones (from the pituitary).

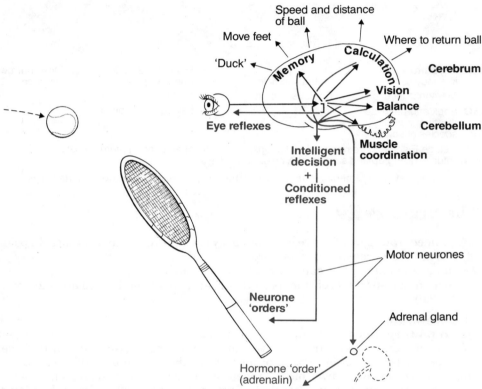

Fig. 12.7 Coordinating role of the brain in returning a tennis ball

12.8 Misused Drugs

Certain substances used by Man affect the nervous system and particularly the brain. These include alcohol, tobacco, misused medicines, e.g. heroin, and a wide range of plant products of no medicinal value, e.g. LSD, marijuana. Their use is frequently commonplace in one country and forbidden or illegal in another. For example, Britain uses alcohol extensively but forbids marijuana smoking; the exact opposite applies in some Middle Eastern moslem countries.

Features of the 'drug scene'

1 All 'drugs' *affect the mind*, distorting judgement or sensations – which is their attraction. They probably affect the way neurones transmit impulses at synapses (Fig. 12.3).

2 They may cause *dependence*, i.e. addiction, if used regularly:
(a) *psychological* dependence – addict *thinks* he needs the drug but the habit can be broken with no ill effects.
(b) *physical* dependence – addict *needs* the drug. Breaking the habit results in illness (withdrawal symptoms), often severe.

3 Regular drug use results in *tolerance*: the body needs ever greater amounts to achieve the same effect. This becomes very expensive and addicts may resort to crime to obtain money.

4 *Addiction* imprisons the mind. Taking the drug becomes an obsession in the worst cases, excluding any possibility of a useful life.

5 *Embryos* of pregnant drug users are forced to share the drugs of their mothers through the placenta.

Motives for drug taking

1 *Social pressure,* particularly at parties: 'I don't want to be called chicken'. 'Friends', not drug pushers, are your worst enemy.

2 *Being daring:* 'I'm not afraid of these so-called dangerous things', and 'I will not get hooked'. With all the evidence around you, why test your courage in *this* way?

3 *Escape:* 'I want to forget my problems'. The problems are still there when you recover – drugs do not solve them. Talk about problems with friends; then take positive action to solve them, if necessary with help.

4 *Creativity:* 'Some poets and writers were addicts and did their best work under the influence'. For every famous drug-taker there are many thousands who are failures.

12.9 Alcohol – Ethanol

Alcohol is a **sedative** (sleep-making) drug, *not* a stimulant.
Benefit: taken in moderation it reduces shyness, improving social contact; relieves stress.
Harm: excessive intake can lead to abusiveness, violence, illness and even death. Regular excessive intake (alcoholism) is an addiction causing long-term health and social problems.

Table 12.2 Units of alcohol

Drink	Single whisky, gin	Glass of sherry, port	Glass of wine	Half pint beer	Gives approx. 15 mg alcohol /100 cm^3 blood if drunk all at once
Alcohol content	40%	20%	8–12% (approximate values)	5%	

Maximum healthy drinking at one session (*not* regularly): men, 8 units; women, 6 units. For people of small body size these figures should be reduced.

1 Effects on behaviour

Table 12.3 Alcohol and its short-term effects

Units drunk	Blood level mg/100 cm^3	General effect	Effect on driving
2	30	Less cautious	Greater chance of accident
3	45	Judgement worse	Reaction time slower, e.g. braking
5	75–80	Not in full control	UK legal limit is 80 mg/100 cm^3 If exceeded (breathalyser test) loss of licence for 1 year, minimum
10	150	Slurred speech. Loss of self-control	Very dangerous driver
13	200	Double vision. Loss of muscle coordination	Incapable
26	400	Unconsciousness. Vomit may be inhaled causing death	—

The effects of a drinking session may last more than a day: headache (hang-over), lack of alertness, greater chance of accident at work. A healthy liver breaks down alcohol, reducing the level in the blood by about 10 mg/100 cm^3 per hour.

2 Effects on health of an alcoholic (long-term)

(*a*) *Liver cirrhosis:* liver shrinking due to death of cells poisoned by alcohol. Liver functions become less efficient (see unit 6.11).
(*b*) *Brain damage:* fewer neurones live; cavities in brain enlarge. This is irreversible.
(*c*) *Heart disease:* pumping action weaker owing to lack of vitamin B$_1$, through a neglected diet.

(*d*) *Babies* born to alcoholic mothers are smaller, less intelligent and disfigured (Foetal Alcoholic Syndrome, FAS) in most cases. Miscarriage is also more likely.

3 Social effects

Excessive *spending* on alcohol means little to spend on food and clothes for the family.

Effects of *violence* may have consequences beyond losing friends, e.g. hospitalization, prison, divorce.

Poor work performance may lead to loss of job.

Alcohol is said to be 'the most expensive drug on the market'.

There are at least 300 000 alcoholics in the UK (1986).

More than half the drivers involved in drunken driving accidents are under 20 years old; about 70% of all road deaths at night involve alcohol.

One in five pedestrians fatally injured on the road had excess alcohol in their blood.

12.10 Endocrine System

A variety of endocrine (ductless) glands discharging their products, hormones, in minute quantities directly into the blood. The pituitary is the 'master gland' controlling the rest (see Fig. 12.8).

Hormones are organic compounds (secreted by endocrine glands in minute quantities into the blood) which affect certain specific body parts or processes. These 'messages':

(a) arrive at the speed blood travels;
(b) have long-lasting effects (hours, days);
(c) control factors in the internal environment needing constant adjustment, e.g. blood sugar level; or processes needing integrated control over a long period, e.g. growth or sexual development.

Glands and their hormones

1 Pituitary

(*a*) **Tropic hormones:** stimulate other endocrine glands, e.g. TSH (thyroid stimulating hormone).
(*b*) **Growth hormone:** promotes growth of muscle, bone (protein synthesis). Deficiency results in a dwarf; excess – a giant.
(*c*) **Antidiuretic hormone** (ADH or vasopressin): water conservation in kidney. Deficiency causes *water diabetes* (see unit 10.5).
(*d*) Secretes many other hormones, including *oxytocin* (ensures contraction of uterus during birth and milk ejection during suckling) and *prolactin* (milk synthesis).

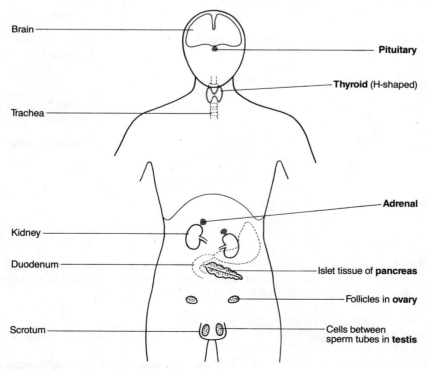

Fig. 12.8 The endocrine system in humans

2 Thyroid

Thyroxine: affects energy release at mitochondria (see unit 1.2) in all cells, raising metabolic rate. Deficiency causes sluggishness, puffy skin; excess produces over-active person with 'pop-eyes'. Deficiency in baby causes *cretinism* – mental and physical retardation (see also mongolism, unit 20.9, and goitre, unit 12.12).

3 Islet tissue of the pancreas

Insulin: causes absorption of glucose from blood into cells, e.g. by liver and muscles to store it as *glycogen*. Deficiency causes sugar diabetes (diabetes mellitus – see unit 10.5).

Glucagon: causes release of glucose into the blood by break-down of glycogen (opposite effect to insulin).

4 Adrenals

Adrenaline (the 'fight or flight hormone'): raises blood glucose level (from glycogen breakdown in liver); increases heart and breathing rates; diverts blood from guts to limb muscles. Nerves (not hormones) stimulate the adrenal, so adrenaline secretion is rapid.

For control of blood glucose level see unit 6.11

5 Ovaries and testes

Produce **sex-hormones,** e.g. oestrogen and testosterone respectively, which promote changes in body proportions, development of gametes and hair, and changes in behaviour and voice, at *puberty* (see Fig. 12.9). (For oestrous cycle, see unit 15.3).

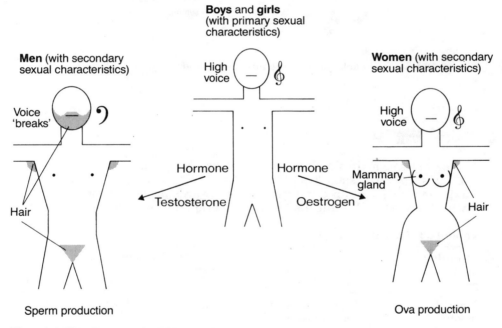

Fig. 12.9 Changes at puberty in humans

12.11 Nervous and Hormonal Systems Compared

Both achieve coordination by *antagonistic action*, e.g. biceps/triceps control of forearm position (see unit 13.9), and insulin/adrenaline control of blood sugar (see unit 6.11).

Table 12.2 Comparison of nervous and endocrine systems

	Nervous system	*Endocrine system*
Speed of 'message'	Fast	Slow
Duration of effect	Short	Long
Precision of 'message'	To a very precise area	A more general effect
Reaction required	Immediate	Long-term

Both systems are *linked to each other*, e.g. hypothalamus (nervous) stimulates the pituitary (hormonal); nerves stimulate adrenal; adrenaline stimulates the heart, just as certain nerves do.

12.12 Feed-Back

Feed-back is the means by which a hormone adjusts its own output by affecting the endocrine glands that cause its secretion. Very important in oestrous cycle (see unit 15.3). If faulty, can cause metabolic disease, e.g. goitre (see Fig. 12.10).

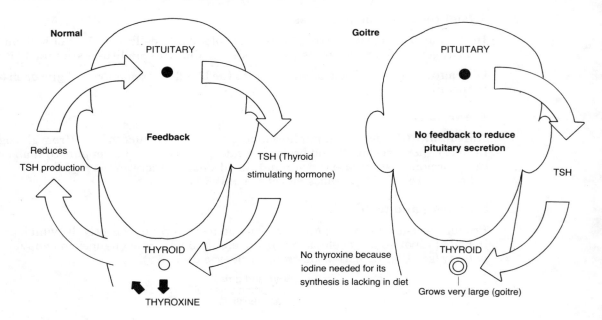

Fig. 12.10 How feed-back controls the pituitary gland

12.13 Taxis

There are three main simple responses to simple stimuli:
   1 taxis, **2** tropism, **3** photoperiodism.
Taxis: movement of an organism bodily towards or away from a stimulus. Applies to many invertebrate animals, unicells and even sperm. For examples see Table 12.3.

Table 12.3 Examples of taxes

Stimulus (and response prefix)	Responses	
	Positive (+ = towards stimulus)	*Negative* (− = away from stimulus)
Light (photo-)	Fly, having escaped swatting, flies towards window	Woodlouse seeks darkness
Water (hydro-)	Woodlouse seeks humid area	
Gravity (geo-)	Fly maggots burrow to pupate	
Chemicals (chemo-)	Blowflies are attracted to meat; so are pond flatworms (*Planaria*)	Earthworms rise from soil dosed with formalin
Contact (thigmo-)	Woodlice huddle together	

Thus woodlice can be described as negatively phototaxic and positively hydrotaxic.

12.14 Tropisms

Tropism: the growth-movement of a plant towards or away from a stimulus. Usually controlled by hormones. For examples see Table 12.4.

Table 12.4 Examples of tropisms

Stimulus	Main shoot response	Main root response	Notes
Light	+ Phototropic	Neutral usually	(Lateral roots and shoots do not behave in this way)
Gravity	− Geotropic	+ Geotropic	

Key: + = positively − = negatively

Phototropism: the growth of plant organs towards or away from light. By growing towards light (positive phototropism), plant shoots get the sunlight they need for photosynthesis. *Mechanism: auxin* (hormone) is made at root and shoot tips (which are sensitive). It diffuses back to region of cell elongation (Fig. 12.11) and here it affects the rate at which cells swell by osmosis (vacuolate). Under normal conditions, equal distribution of auxin gives even growth. With a one-sided stimulus, distribution becomes unequal giving unequal growth.

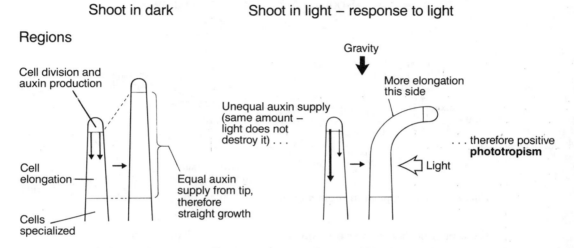

Fig. 12.11 Mechanism for the response of shoots to light from one side.

Experiment to discover the region of sensitivity to light from one side in oat coleoptiles, and their response

Using *many* coleoptiles, grown in the dark, shields of three kinds are applied (see Fig. 12.12). The seedlings are now given light from one side (only one seedling of each type shown):

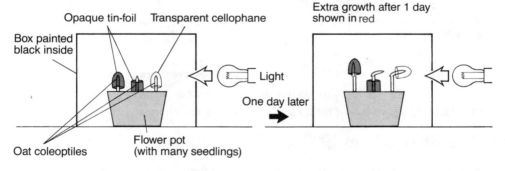

Fig. 12.12. The response of oat coleoptiles to light from one side

It may be concluded that (*a*) the coleoptile tip is sensitive to light;
(*b*) the response is positively phototropic.

12.15 Geotropism

Geotropism: the growth of plant organs towards or away from gravity. By growing downwards (positive geotropism) plant roots usually find the water they need. Plant shoots grow upwards (negative geotropism), so finding light.

Mechanism: auxin diffuses back from the tip unequally. In the root the greater concentration on the lower side *inhibits* growth there and the root grows faster on the upper side. In the shoot the greater concentration of auxin *stimulates* growth (as it does in phototropism). (See Fig. 12.13.)

Shoot in dark – response to gravity

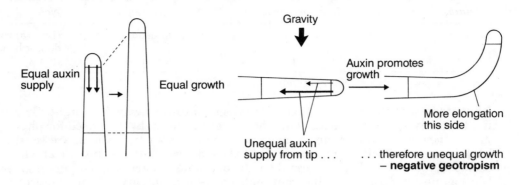

Root in dark – response to gravity

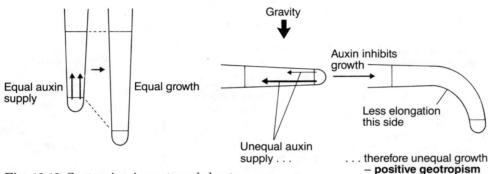

Fig. 12.13 Geotropism in roots and shoots

Experiment to test the response to gravity of bean roots

Five beans are pinned to each of two klinostats (one only shown).

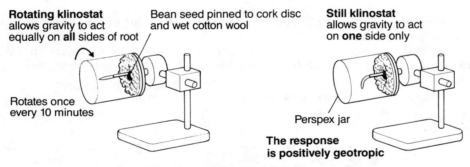

Resulting growth after 1 day shown in red

Fig. 12.14 The response of bean roots to gravity from one side

12.16 Photoperiodism

Photoperiodism: an organism's response to *length of day* (or night) by initiating an important event in the life cycle. This event (e.g. migration of birds, emergence of some insects from pupae, or flowering in most angiosperms) is usually linked with reproduction, ensuring that it occurs in the right season.

1 Angiosperms

Length of night is 'measured' by the leaves, within which is a blue pigment, *phytochrome*, which reacts differently to day and night. This acts as the 'clock' to start synthesis of a flowering-hormone (*florigen*). Florigen starts flower formation.

2 Mammals

Many *mammals* have a breeding season, e.g. deer, lions (but not Man). Day-length probably influences the pituitary gland via the eyes and brain. The pituitary secretes hormones influencing the testes and ovaries to grow and produce gametes. After breeding, the testes and ovaries become small again.

Daily rhythm probably sets Man's biological clock – which requires re-setting, after long journeys by jet, to new rhythms of day and night (jet-lag).

13 SUPPORT AND LOCOMOTION

13.1 Principles of Support

The mass of an organism is supported by its environment (water, land or air).

Plants transmit their weight to it via *cell walls*; and animals via their *skeletons* (they have no cell walls). Since most animals move, their skeletons are used both for *support and locomotion*. The environment provides support by a '*buoyancy*' effect too (Table 13.1).

Table 13.1

Water	Land and Air
Great support: organism is made lighter by the mass of water it displaces	Very *little support* from air since volume of air displaced has small mass
Therefore plants and animals only need relatively *weak 'skeletons'*	Therefore *strong 'skeletons'* needed, particularly if organisms are large

13.2 Support in Plants

Flowering plants use cell walls to support their weight. Cellulose walls are weak; woody ones strong.

1 **Small plants** (herbaceous), e.g. grasses, use osmosis to inflate vacuoles with water (see unit 7.4). The vacuoles press outwards on the cytoplasm, causing the cells to inflate (become turgid). This stiffens cell walls. The combined effect of many such turgid cells causes stems to stiffen (the experiment in Fig. 7.5 demonstrates this).
   Woody cells play a small part in support.
2 **Larger plants** (shrubs, trees) use woody cells to provide most of their support. These cells are fibres and xylem vessels (see unit 7.9). Both kinds of cell are dead, so osmosis plays no part in their support.
3 **Water plants,** e.g. Canadian pond-weed, are largely supported by the water outside them. There is little xylem so stems are weak.

13.3 Support and Locomotion in Animals

The skeleton is used (a) to support, (b) for locomotion – where *shape* of the body and limbs is important (see Fig. 13.1).

Water	Land	Air
(a) *Weak skeleton* (sharks manage on cartilage) and massive animals (e.g. whales and giant squids) are possible	*Strong skeleton* essential because full weight of body acts through the small areas where limbs are attached to body. Also prevents internal organs crushing each other as they sag downwards	
(b) *Streamlining* and *buoyancy* important in saving energy when moving through water (dense medium). See unit 21.16	*Foot design* important for efficient movement on, e.g. sand (camel), rock (mountain goat), trees (leopard)	*Streamlining* and *wing design* important for sufficient lift and speed (see unit 21.18)

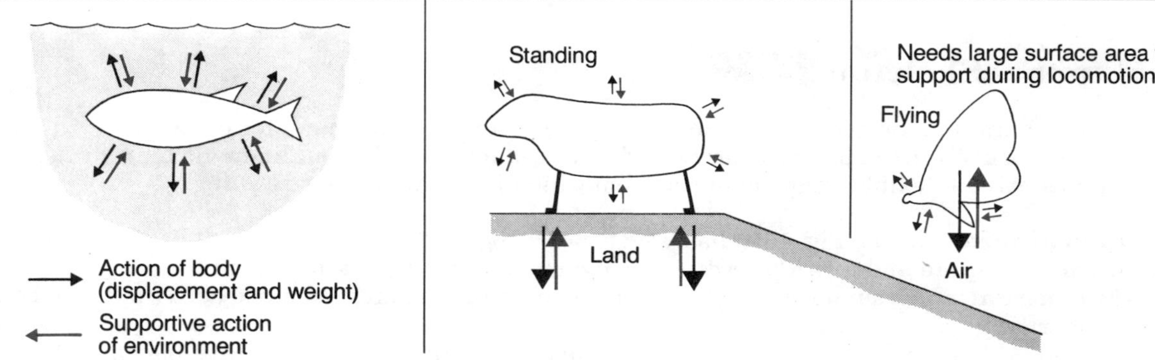

Standing

Needs large surface area for support during locomotion

Flying

Land

Air

→ Action of body (displacement and weight)

← Supportive action of environment

Fig. 13.1
The influence of environment on skeleton design

13.4 Principles of Movement

1 **Muscle** can only *contract* (pull) – cannot push. To be lengthened again it must (*a*) relax, (*b*) be pulled back into shape – by another muscle, its antagonist, e.g. biceps and triceps (see Fig. 13.8). Thus muscles work in *antagonistic pairs*.

2 **Nerve impulses** are essential to make muscles *contract* (except heart). The antagonistic muscles are kept *relaxed* by impulses too (reflex inhibition).

3 **Skeleton** transmits the contraction force of muscle to the environment, e.g. water, land, air during swimming, walking and flying.

4 **Load-bearing surface** in contact with the environment must get purchase on it if locomotion is to result, e.g. fish tail on water, bird wing on air, hooves on ground or claws on trees.

13.5 Mammal Tissues for Support and Locomotion

The skeleton is mainly bone. Bone is covered at joints by cartilage. Ligaments connect bones to bones. Tendons connect muscles to bones. Each tissue has its own special properties:

1 **Bone** is both hard and flexible to some extent. *Bone cells*, arranged in cylindrical layers, secrete the mineral calcium phosphate to give hardness. Cylinders are strong. Bone cells are attached to a network of fibres which give flexibility (see Fig. 13.2).

Soak a small long-bone in 3% hydrochloric acid for three days. It comes out rubbery – the minerals have been dissolved.

Heat a bone in a Bunsen burner flame. It becomes brittle and breaks easily – the fibres have burnt away.

2 **Cartilage** is a rubbery protein secreted by cells. It cushions the ends of bones at joints (shock absorber and smooth surface) – see Fig. 13.2.

3 **Connective tissue** is of two kinds:
(*a*) **Ligaments** are elastic fibres allowing 'give' at joints and between vertebrae.
(*b*) **Tendons** are inelastic fibres. They ensure that muscles pull bones immediately, without having to take up 'slack'.

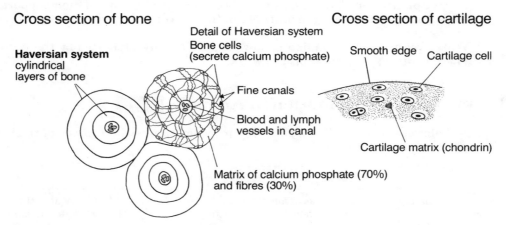

Fig. 13.2 Structure of bone and cartilage

4 **Muscle:** cells containing protein that contracts.
(*a*) **Involuntary muscle,** e.g. in gut, causing peristalsis (Fig. 6.10); in iris, affecting pupil size; and arteries, affecting blood flow (Table 8.2). None are controlled by will-power.
(*b*) **Voluntary muscle,** e.g. in arm. Controlled by decision.

13.6 Mammal Skeleton

1 **Skull:** cranium protects brain; houses all major sense organs; jaws for chewing.

2 **Vertebral column:** protects nerve cord; acts as anchorage for four limbs via limb girdles and for ribs. Also a flexible, segmented rod from which internal organs are slung.

Typical vertebra (see Fig. 13.4) has:
(*a*) **neural spine** and *lateral processes* for ligament and muscle attachment;
(*b*) **centrum** which supports weight (aided by 'discs') and makes red blood cells (in its red bone marrow);
(*c*) **neural canal** which houses and protects nerve cord (nerves exit via two adjacent notches).

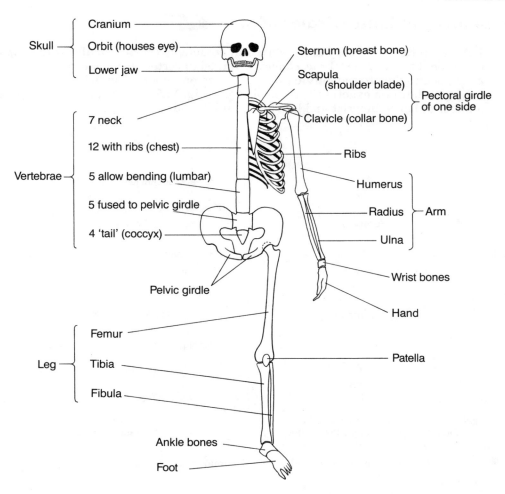

Fig. 13.3 Main parts of the human skeleton

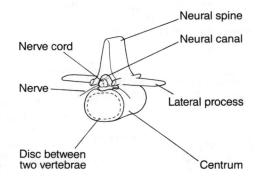

Fig. 13.4 A generalized vertebra (with nearby structures shown in red)

Discs are shock-absorbers between vertebrae. Their tough fibrous coat of connective tissue encloses a pulpy centre. A 'slipped disc' occurs when excessive pressure causes the disc to *bulge*, pressing on a nerve and causing pain. Most often, lifting heavy objects with a bent back causes disc damage – particularly in the lumbar region (small of the back) – see Fig. 13.3

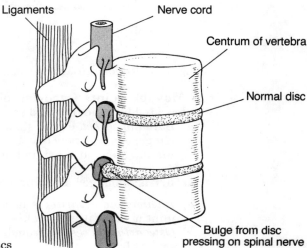

Fig. 13.5 Normal and 'slipped' intervertebral discs

13.7 Limbs and Limb Girdles of Man

1 Arm is attached loosely to the vertebral column by the **pectoral girdle;** *scapula* has muscles to attach it to chest vertebrae; *clavicle* is linked to vertebrae via sternum and ribs (see Fig. 13.3).
2 Leg is attached to the **pelvic girdle**. This strong hoop of bone is fused firmly to 5 vertebrae.
3 Limbs: built on exactly the same plan (see Fig. 13.6) – one upper bone, two lower, and same number of bones in wrist and hand as in ankle and foot.

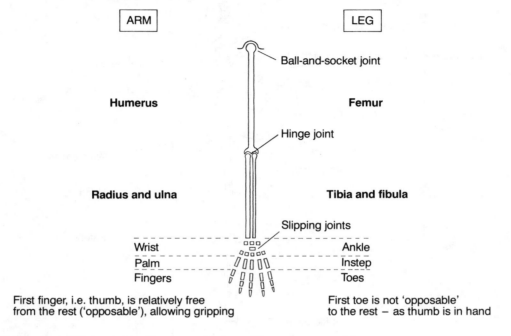

Fig. 13.6 Comparison of bones and joints of human arm and leg

13.8 Joints

Joints are where bones are linked (see Fig. 13.7).
1 Immovable joints (sutures): wavy interlocking edges of bone, held together by connective tissue, e.g. bones of cranium.

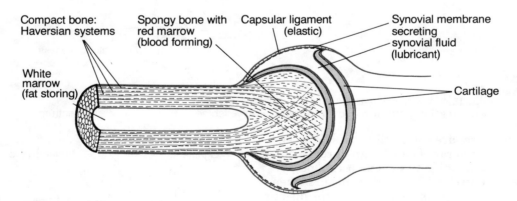

Fig. 13.7 Section through mammal synovial joint and bone

2 Movable joints (synovial joints): bones have cartilage ends; these move on each other, lubricated by synovial fluid secreted by a synovial membrane within a joint capsule. Types:
(*a*) *Ball-and-socket*, e.g. at shoulder, hip – rotation in *two* planes of space (see Fig. 13.6).
(*b*) *Hinge*, e.g. at elbow (see Fig. 13.8) and knee – movement in *one* plane only (like door).
(*c*) *Slipping*, e.g. at wrist and ankle – limited rocking movement.

Arthritis: damaged and painful joints which swell.

(*a*) *Rheumatoid arthritis* is usually a hereditary disease. Connective tissue grows across the joint making it immovable.
(*b*) *Osteoarthritis* results from break-down of cartilage through old age. Joints no longer move smoothly.

Artificial joints of titanium alloys (metal) and nylon may be inserted surgically to replace arthritic joints.

13.9 Movement of an Arm

Bending: nerve impulses make the biceps contract, so raising the forearm. Other nerve impulses, travelling to the triceps, *inhibit* its contraction, so it is extended by the biceps through leverage.
Extending: nerve impulses cause contraction of the triceps. This extends the biceps, and the arm moves down.
The biceps and triceps muscles are *antagonists*. Nerve impulses stimulate one muscle while other impulses inhibit the other muscle when movement occurs.

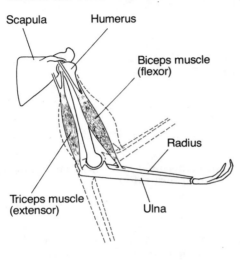

Fig. 13.8 Movement of the forearm in Man (hinge joint)

13.10 Functions of Mammal Skeletons

1 **Support:** of body off ground; of internal organs, preventing crushing.
2 **Shape:** important adaptations, e.g. Man's hand, bat's wing, porpoise's streamline and flippers.
3 **Locomotion:** system of levers.
4 **Protection:** cranium protects brain; ribs protect heart and lungs.
5 **Breathing:** role of ribs (see unit 9.8).
6 **Making blood cells:** in red bone marrow, e.g. of ribs, vertebrae (and see Fig. 13.7).
7 **Sound conduction:** three ossicles in middle ear (see unit 11.4).

13.11 Sports Injuries

These may result from
1 lack of training – body not prepared for strains, e.g. pulled muscles;
2 sudden excessive stress, e.g. fractured bones, knee joint damage;
3 over-use – injuries never given time to heal properly;
4 self-inflicted causes.

1 Lack of training
(a) *Stamina lacking:* the body is most vulnerable when tired. Most rugby injuries occur early in the season and in the last quarter hour.
(b) *Strength lacking: both* antagonistic muscles must be built up and on both sides of the body to avoid self-injury.
(c) *Flexibility lacking:* 'warming up' must include stretching, to prepare tendons and ligaments for stress. Vital in sprinters.
(d) *Skill lacking:* poor positioning and timing can result in collision or other sudden stress.

2 Sudden stress
(a) *Fractures:* bones may be cracked or broken across. The bone must be immobilized, e.g. by splints, in the right position by experts. This allows
(i) broken ends to be joined by connective tissue;
(ii) bone cells to attach to fibres and secrete bone minerals;

(iii) reabsorption of any extra bone over some months.
(b) *Knee joint injury*, e.g. in football and squash (see Fig. 13.9).

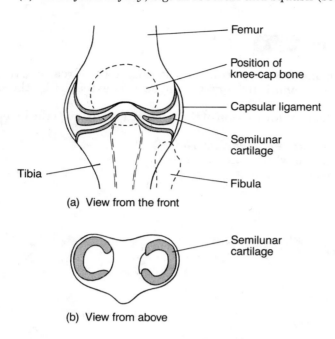

(a) View from the front

Semilunar cartilage

(b) View from above

Fig. 13.9 Diagram of a knee joint (severe twisting of this hinge joint may lead to splitting of the semilunar cartilages – which then have to be removed surgically)

3 Over use
Damage due to severe competition must be given time to heal *fully*. Rest is needed to avoid 'staleness' and a drop in performance that may lead to obvious injury.

4 Self-inflicted causes
(a) Contact sports, e.g. rugby, need organization on a weight basis (and not age basis) at junior level.
(b) Anabolic steroids, while improving performance by building up muscles, have serious consequences. These include increases in atherosclerosis (see unit 9.11) and blood pressure; damage to the liver; and male sterility – when used long-term.

14 REPRODUCTION: MAINLY PLANTS

14.1 Asexual and Sexual Reproduction Compared

No individual organism is immortal; reproduction avoids extinction. Most organisms reproduce sexually, many asexually as well.

Table 14.1 Comparison of asexual and sexual reproduction

	Asexual	*Sexual*
Parents	One	Two (unless parent is hermaphrodite, i.e. both sexes in same individual)
Method	Mitosis forms either: (a) reproductive bodies, e.g. spores, tubers, or (b) replicas of adult by outgrowth, e.g. runners	Meiosis forms gametes (sperm and ova) which fuse to form zygotes (at fertilization) Zygote grows by mitosis into new organism

	Asexual	*Sexual*
Offspring	Genetically identical to parent	Not identical – half its genes are maternal (mother's), half paternal (father's)
Advantage	Maintains a good strain exactly	Produces new varieties which, if 'better', favour survival and in the long-term their evolution (see unit 18.3)
Disadvantage	Species liable to be wiped out, e.g. by disease, if not resistant to it	Excellent individuals, e.g. prize milk-cow, cannot give identical offspring
Other points	Only one arrival needed to colonize a new area Often more rapid than sexual methods Always increases population	Both sexes needed Not rapid Need not increase population (two parents may produce only one offspring, then die)
Occurrence	Very common among plants and *simple* animals, e.g. *Amoeba*	Almost all plants and animals

14.2 Asexual Methods of Reproduction

All the offspring from one asexually-reproducing parent are known as a **clone** (a genetically identical population).

1 Binary fission, e.g. bacteria (see unit 3.2), *Amoeba* (see unit 3.11).

2 Spores, e.g. fungi (see unit 3.7), mosses (see unit 21.2).

3 Budding, e.g. tapeworm (see unit 21.15).

4 Identical twinning, e.g. in humans, a single zygote may develop into two babies.

5 Vegetative propagation by outgrowths of new plantlets usually from *stems* (see Fig. 14.1) but sometimes from *leaves*, e.g. *Bryophyllum*. Many of these methods of asexual reproduction also achieve **perennation** (survival over winter in a dormant state).

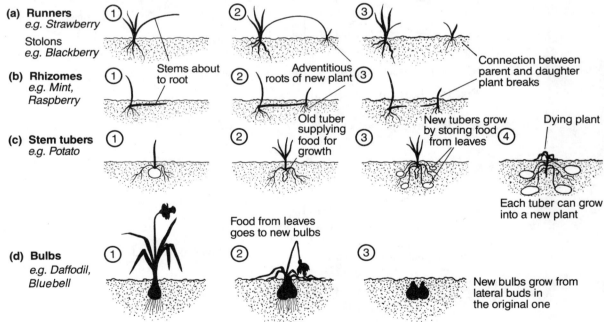

Note: all the plants named also reproduce sexually (by flowers) at the end of the growing season.

Fig. 14.1 Methods of asexual reproduction (and perennation) in flowering plants

Potato tuber: an underground *stem*-tip swollen with food (especially starch) received from the parent plant, which dies down in autumn. Each tuber is a potential new plant (thus *asexual reproduction*) and allows *perennation*. New shoots and adventitious roots arise from 'eyes' (see Fig. 14.2) in spring.

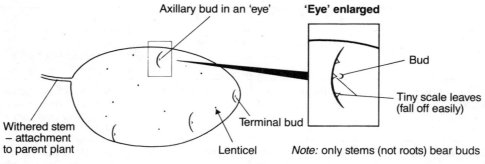

Fig. 14.2 A potato tuber

14.3 Surviving Winter – Perennation

Simple Plants may survive as spores, e.g. fungi, mosses. Some survive below ground after the foliage dies down, e.g. ferns.

Flowering plants

(a) **Annuals**, e.g. poppies, complete their life cycle in one year, surviving as dormant seeds. Cold is often needed to break dormancy, i.e. to make germination possible next spring.

(b) **Biennials** Most root crops, e.g. carrots, live for two years. They survive underground at end of year 1, and use their stored food to flower and make seeds in year 2.

(c) **Perennials** live for a number of years, surviving both as seed and as vegetative structures (see Fig. 14.1).

Evergreens either partially die down, e.g. grasses, or have leaves that resist damage by frost, snow and wind, e.g. holly.

Deciduous plants shed their leaves in autumn (see Fig. 14.3). This reduces wind resistance and so avoids uprooting in gales. It also reduces transpiration at a time when water (frozen) may not be available to the plant.

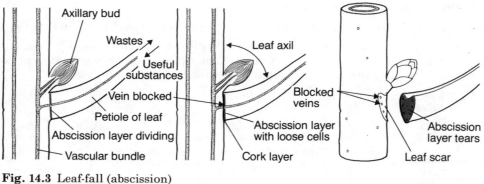

Fig. 14.3 Leaf-fall (abscission)

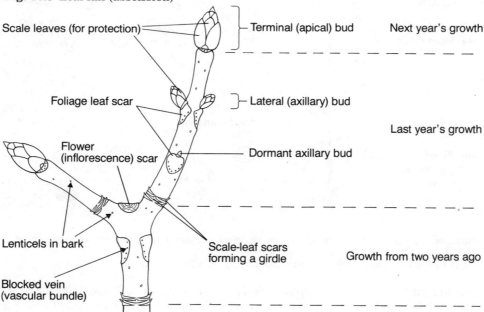

Fig. 14.4 A winter twig

Winter twigs (see Fig. 14.4) protect next year's leaves and flowers within *buds*. When these organs expand in spring, their protecting leathery leaves drop off, leaving *girdle scars* – marking the beginning of a year's growth. If a flower (inflorescence) grows from the bud, growth will end there when it withers, leaving a scar. *Lateral buds* take over growth. *Dormant buds* (small) lie in reserve, only growing if other buds die. *Terminal buds* normally continue growth.

In autumn the green stem becomes covered by protective *cork* and its stomata are replaced by lenticels (see Fig. 14.5) to allow gaseous exchange. *Leaf scars* form when leaves fall – a process controlled by hormones.

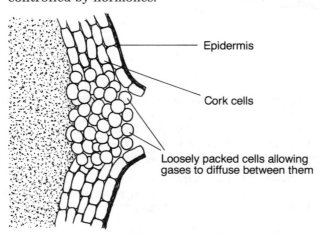

Fig. 14.5 Section through a lenticel in a stem

1 Leaf dying	2 Stem sealed off from petiole	3 Leaf-fall
Useful substances passed into stem. Wastes, e.g. tannins, received by leaf for excretion. New layer of dividing cells (abscission layer) arises.	Veins blocked, therefore leaf dries up. Abscission layer forms cork cells on stem-side; loosely packed cells (line of weakness) on leaf-side.	Wind or frost tears leaf off at abscission layer, leaving leaf scar.

14.4 Vegetative Propagation

Artificial methods of asexual reproduction are used by Man to
(*a*) maintain good varieties of house-plants and some crop plants;
(*b*) multiply rapidly new varieties which arise by mutation (see unit 17.15);
(*c*) multiply rapidly new varieties produced by selective breeding (see unit 18.6);
(*d*) maintain seedless oranges and grapes – no other way possible.

A Cuttings: lengths of stem, e.g. *Pelargonium*, or a leaf, e.g. African violet, made to grow into complete plants. This method requires:
(*a*) *sufficiently large piece:* enough food reserves to form the missing roots;
(*b*) *sand/peat mix for rooting:* enough air for respiration where roots are forming; enough water to supply needs;
(*c*) *removal of most leaves*: to reduce transpiration (plant would dry out);
(*d*) *rooting hormone:* applied to cut end, starts cells dividing to form roots.

B Grafting: the insertion of a shoot or bud onto a related plant (see Fig. 14.6a and b).
The two grow into one plant which has the advantages of:
(*a*) the *vigour* of a specially chosen root – the **stock**;
(*b*) the *quality* of the product (flowers or fruit) on the grafted shoot – the **scion**.

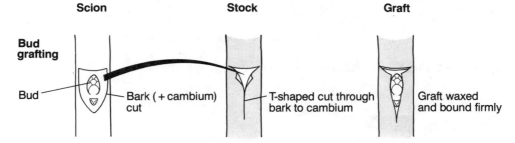

Fig. 14.6(a) Bud grafting

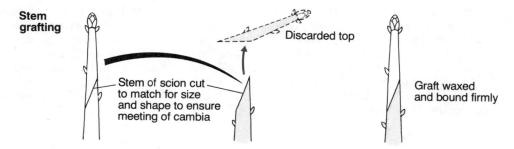

Fig. 14.6(b) Stem grafting

Grafting requires:

(*a*) that cambia (see unit 7.9) of stock and scion must meet (to grow together);
(*b*) firm binding at junction (to prevent joining tissue tearing);
(*c*) autumn grafting (to minimize death from excessive transpiration);
(*d*) waterproofing the cuts with tape or wax (minimizes infection and desiccation);
(*e*) compatible species (lemon will not graft onto an apple).

Note: Neither stock nor scion are altered *genetically* by grafting. Thus in a garden rose, 'Masquerade', large beautiful roses form on the scion, but only small wild briar roses can form on 'suckers' (stems) sprouting from the stock. The 'Masquerade' genes do not affect briar genes – or vice versa.

Grafting is used extensively in producing grapes, apples, pears, citrus fruit and roses. 'Family' apple trees have a number of different varieties of apple growing on the same trunk, from separate grafts.

C Tissue culture: the growth of whole plants from small groups of cells using growth media and hormones.

Advantages: *very large numbers* of plants can be grown commercially from a single one in a short time – much faster than by cuttings.

Method:
1 Tissue, e.g. pith, is scraped out of the parent plant. It is spread on sterile agar containing nutrients (sucrose, mineral salts, vitamins) and auxin in a petri dish.
2 After some weeks each group of cells has divided into a formless mass of many thousands of similar cells – a **callus**.
3 The callus is now made to grow roots, stem and leaves by a special mix of hormones in agar containing nutrients.
4 The resulting small plants can be planted out (see Fig. 14.7).

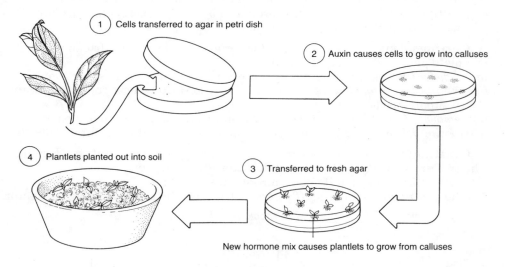

Fig. 14.7 Plant tissue culture to form new plants

14.5 Flowers

A flower is the organ of sexual reproduction in flowering plants (see Fig. 14.8). It is usually bi-sexual (hermaphrodite) but sometimes unisexual, e.g. holly.

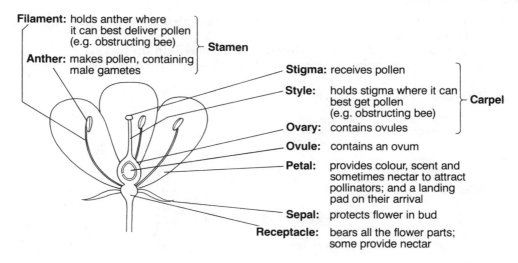

Filament: holds anther where it can best deliver pollen (e.g. obstructing bee)
Anther: makes pollen, containing male gametes
— **Stamen**

Stigma: receives pollen
Style: holds stigma where it can best get pollen (e.g. obstructing bee)
Ovary: contains ovules
— **Carpel**

Ovule: contains an ovum
Petal: provides colour, scent and sometimes nectar to attract pollinators; and a landing pad on their arrival
Sepal: protects flower in bud
Receptacle: bears all the flower parts; some provide nectar

Fig. 14.8 Structure and functions of the parts of a generalized insect-pollinated flower

A flower consists of an expanded stem-tip, the **receptacle,** on which is borne four rings of modified leaves:
(i) *sepals* – almost leaf-like but protective
(ii) *petals* – often coloured and attractive
(iii) *stamens* – male parts
(iv) *carpels* – female parts
There are **two main stages in sexual reproduction:**

1 Pollination: transfer of pollen from stamens to stigmas.

2 Fertilization: fusion of male gamete with female gamete inside the ovule. This results from the growth of pollen tubes from the pollen on the stigmas to the ovules.

14.6 Self- and Cross-pollination

Self-pollination: transfer of pollen from any stamen to any stigma on the *same plant* (not necessarily the same flower). Results in fewer varieties of offspring than cross-pollination. Frequent in cereal crops, grasses.

Cross-pollination: transfer of pollen of one plant to the stigmas of *another plant of the same species*. Thus rose pollen landing on an apple stigma will *not* germinate there. Results in a great variety of offspring. Since variety assists survival, many plants have means of improving the chances of cross-pollination (see Fig. 14.9).

14.7 Wind and Insect Pollination

Table 14.2 Comparison of flowers adapted for wind or insect pollination

	Wind pollination	*Insect pollination*
1 *Petals*	**Not attractive:** usually green, unscented; no nectar **Small:** leaving stamens and carpels exposed	**Attractive:** coloured, scented, often with nectaries **Large:** protect stamens and carpels inside
2 *Stamens*	Long filaments and large mobile anthers **exposed to wind**	Stiff filaments and anthers **obstruct visiting insects**
3 *Pollen*	**Large quantities** (enormous chances against it all reaching stigmas). Small, dry, light (easily wind-borne)	**Small quantities** (more certain 'delivery service'). Rougher, sometimes sticky (to catch on insect 'hairs')
4 *Stigmas*	**Large, exposed** to wind (to catch passing pollen)	**Small, unexposed,** sticky with stiff style (to obstruct insects)
Examples	Plantain, grasses, hazel, oak	Buttercup, deadnettle, horse-chestnut, cherry

Note: certain flowers, which appear to be suited for insect pollination, in fact use other methods. For example:
(*a*) *peas and French beans* **self-pollinate** when still in the bud stage;
(*b*) *dandelions* develop seed from ovules without fertilization, i.e. asexually.

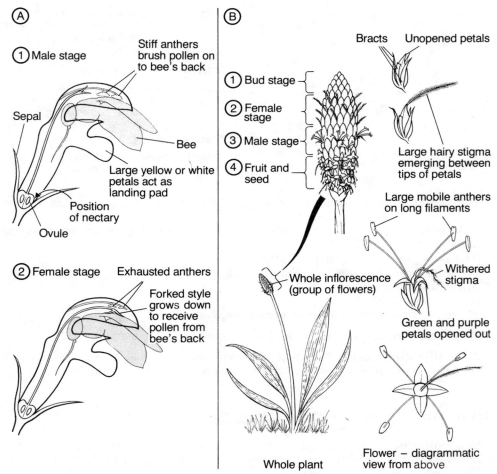

Fig. 14.9 (A) Features of an insect-pollinated flower illustrated by the deadnettle (seen in section) (B) Features of a wind-pollinated flower illustrated by the narrow-leaved plantain

14.8 Fertilization and its Consequences

1 A pollen grain of the right kind on the stigma germinates to form a pollen tube (see Fig. 14.11).
2 The pollen tube grows down the style and ovary wall to the micropyle of the ovule.
3 A male nucleus passes from the pollen tube into the ovule to fuse with the ovum (fertilization):
   male nucleus + ovum $\longrightarrow$ zygote cell
4 The zygote divides by mitosis (see unit 17.10) to form the **embryo:**
   zygote $\longrightarrow$ plumule (shoot), radicle (root) and cotyledons (seed leaves)
5 The integuments (thin wrappings around the ovule) grow and harden into the **testa** (seed coat), still with its micropyle.
   Thus embryo + testa = **seed** (see unit 16.4).
6 The ovary wall grows into the **fruit** – which contains seed(s).
7 Most of the other flower parts drop off, i.e. petals, stamens, stigma and style, and often sepals too.

14.9 From Flower to Seed

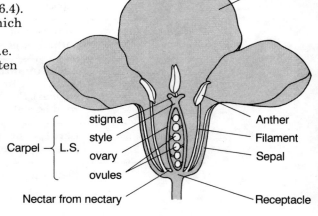

Fig. 14.10 Wallflower flower with a sectioned ovary and a petal, sepal and stamen removed

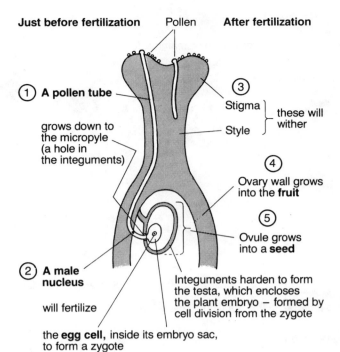

Fig. 14.11 Fertilization and its consequences in a wallflower flower

14.10 Fruits

Fruits serve two main functions:

1 Protection of seed: particularly important when fruit is eaten by animals, e.g. in stone-fruits – peaches, cherries: inner part of fruit is hard.

2 Dispersal of seed (see Fig. 14.12).

(*a*) *Avoids overcrowding* (more likely with some methods of asexual reproduction, e.g. runners). Seedlings do not have to compete with parent for light, water and mineral salts.

(*b*) *Helps colonization* of new areas.

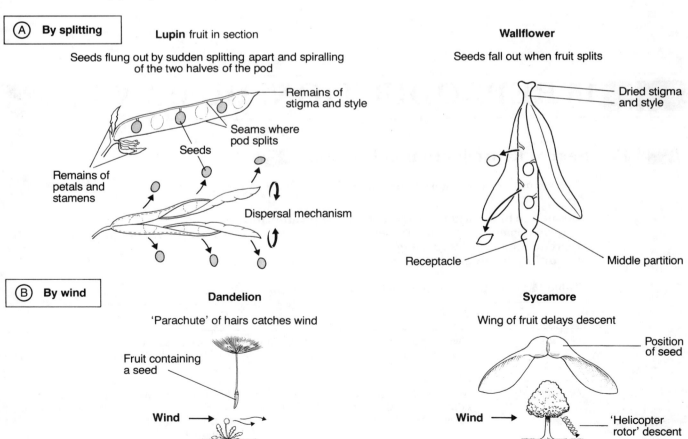

Fig. 14.12 Dispersal of seed by fruits (continued on p. 98)

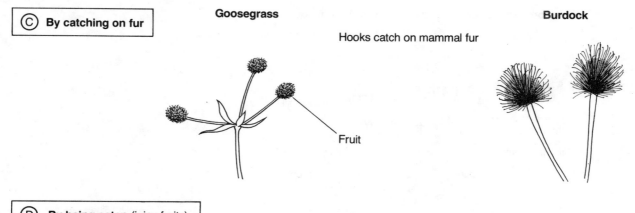

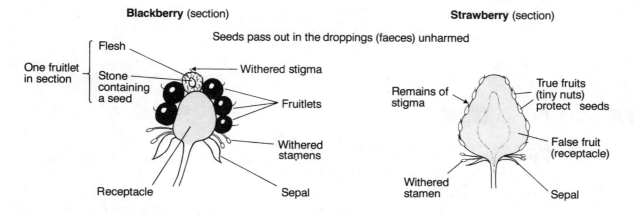

Fig. 14.12 (*contd.*)

15 REPRODUCTION: HUMANS

15.1 Sexual Reproduction in Humans

The sexual organs of man and woman are shown in Fig. 15.1.

Sequence of events in human sexual reproduction
1 Development of secondary sexual characteristics at **puberty** (12–14 years old) making reproduction possible (see Fig. 12.9).
2 **Gamete production**

Table 15.1   Comparison of gamete production in humans

	Male	*Female*
Gonads	Two **testes**, kept outside the body in a sac (scrotum), produce sperm	Two **ovaries**, kept within the body cavity attached to a ligament, produce ova
Gametes	Many millions of **sperm** formed continuously throughout life after puberty (see Fig. 15.2)	Many thousands of potential **ova** formed before birth, but only about 400 will be shed between puberty and menopause (about 45 years old: end of reproductive life)

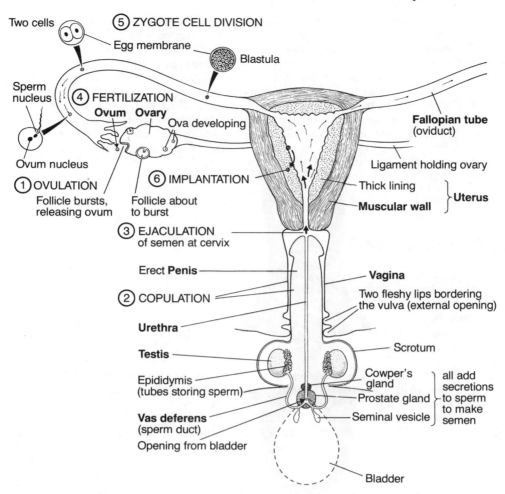

Fig. 15.1 Human male and female sex organs at copulation and the events leading to implantation

Table 15.1 (*contd.*)

	Male	*Female*
Gamete release	About **200 million** sperm are ejaculated into female by *reflex action* of the penis during copulation. They pass along sperm duct and urethra, picking up nutritive secretions from glands to form *semen*.	Usually only **one** ovum is shed *automatically* per month (unit 15.3) from an ovary. It passes into the oviduct (Fallopian tube), the only place where it can be fertilized. Once in the uterus, the ovum is lost.

3 Copulation: the erect penis transfers sperm during ejaculation from the testes of the male to the end of the vagina (cervix) of the female.

4 Fertilization: any sperm that manages to swim into an oviduct containing an ovum has a chance of fertilizing it. One sperm only enters the ovum and the two nuclei fuse, forming the zygote cell.

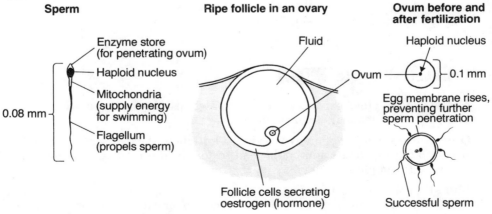

Fig. 15.2 Human gametes and fertilization

5 Cell division: the zygote divides into a ball of cells (blastula); passes down oviduct.

6 Implantation: the blastula sinks into the uterus lining.

7 Growth: the blastula grows into two parts – the **embryo** and its **placenta**, joined by the umbilical cord. The embryo lies within an **amnion**, a water-bag, which cushions it from damaging blows and supports it. Growth lasts 38 weeks (9 months) – the **gestation** period. Premature birth, before 7 months, results in the embryo's death (spontaneous abortion or miscarriage).

8 Birth: the **baby** is pushed head first through a widened cervix when the uterus muscles contract. This bursts the amnion. The umbilical cord is cut by the midwife. When the baby's end of the cord dries up, it drops off leaving a scar (the navel). Babies are usually born head first.

A baby about to be born *feet* first poses difficulties. The doctor may get it out by cutting open the abdomen and uterus (Caesarian section).

9 After-birth: within 30 minutes after the birth, further contractions of the uterus expel the **placenta**.

15.2 Placenta

The placenta: a temporary organ grown in the uterus during gestation to supply the needs of the embryo. These needs are:

Supply of:
(*a*) *food* – soluble nutrients, e.g. amino-acids, glucose.
(*b*) *oxygen* – for respiration.

Removal of:
(*a*) *urea* and other wastes.
(*b*) *carbon dioxide*.

This exchange of substances occurs at capillaries, inside villi, at the end of the umbilical cord (see Fig. 15.3). The villi lie in spaces filled with mother's blood. Mother's blood does *not* mix with the embryo's blood. If it did
(*a*) her blood pressure might burst the embryo's blood vessels;
(*b*) blood clumping might occur if the two blood groups were different (see unit 17.3).

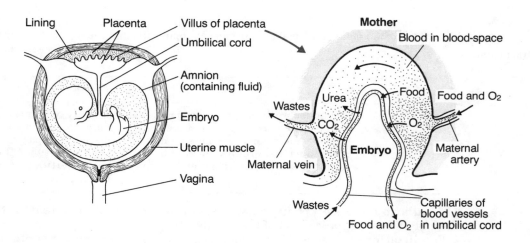

Fig. 15.3 Placenta – relationship between mother and embryo

15.3 Menstrual Cycle

Menstrual cycles: periods of approximately 28 days during which a reproductive woman alternately ovulates and menstruates.

Ovulation: shedding of an ovum when a follicle in the ovary bursts (see Fig. 15.4). Copulation within 3 days of ovulation could lead to fertilization, so the uterus lining (endometrium) is prepared for implantation (see Fig. 15.4).

Menstruation: shedding of most of the uterus lining 14 days after ovulation, when fertilization or implantation are unsuccessful. This occurs over 4 days as a loss of up to 500 cm^3 of blood and tissue through the vagina.

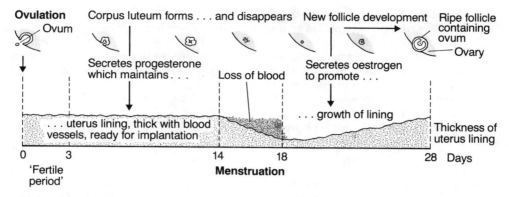

Fig. 15.4 The menstrual cycle

15.4 Contraception

Contraception: the prevention of fertilization and implantation.

1 Unreliable methods
(a) *Withdrawal:* removing penis just before ejaculation;
(b) *Rhythm:* not copulating during the 'fertile period' of the menstrual cycle, i.e. when an ovum is passing down the oviduct (see Fig. 15.4). Ovulation is sometimes irregular; and sperm may survive 48 hours inside the woman.

2 Reliable methods (see Fig. 15.5) include:
(a) *temporary* methods allowing sensible spacing of a family. This places less physical strain on the mother; and more time and finance can be given to the care and attention of each child.
(b) *permanent* methods, when desired family size has been reached. Removal of testes (castration) or ovaries achieves the same result but is undesirable since a person's 'nature' is changed owing to lack of sex hormones from these organs.

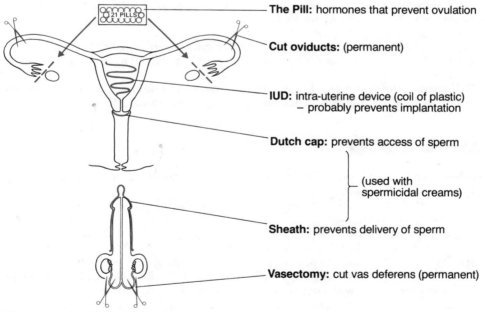

Fig. 15.5 Reliable methods of contraception

Practice of contraception world-wide is essential if humans are to avoid destruction of their environment by pollution (see unit 20.6), erosion, and social problems. A stable or falling birth-rate has been achieved in a number of industrialized nations already; developing nations lag behind in effective contraception. A notable exception is the People's Republic of China where there are severe social penalties for couples who have more than one child. The birth rate has fallen (see unit 16.9).

15.5 Sexually Transmitted Diseases (STD)

Sexually transmitted diseases are those passed on through copulation.

Copulation can occur in two ways:

1 *Heterosexually* (between man and woman) – the normal way, used in reproduction (see unit 15.1). Both penis and vagina are adaptations for this biologically important act.

2 *Homosexually* (between man and man). A penis is inserted into his partner's rectum. The rectum is not adapted for the sex act and tears easily, causing bleeding. This act has no biological function, so some people regard it as abnormal.

Bisexuals copulate both heterosexually and homosexually.

There are two main types of pathogen in STD:

1 Bacteria – which can be killed by antibiotics;

2 Viruses – incurable.

BACTERIA

1 Syphilis: this disease occurs in three stages; by the third it is incurable.

(*a*) A painless *sore* appears at the point of contact, e.g. the penis tip or the cervix (thus the woman is often unaware of her infection) within 90 days of sexual contact. This disappears.

(*b*) Four to eight weeks later skin *rashes* may appear or patches of hair may fall out.

(*c*) After some weeks, infection reaches the *nervous system*, leading to paralysis, idiocy, blindness, etc.

Unborn babies can be infected via the placenta; they may suffer abnormalities or be born dead.

2 Gonorrhoea: within two to eight days of sexual contact a yellowish discharge of mucus may appear from penis or vagina. In both sexes there may be permanent *problems with urination*; and both may become sexually *sterile* (through blocking of the sperm ducts or the oviducts with scar tissue where there has been infection).

VIRUSES

1 AIDS (Acquired **I**mmune **D**eficiency **S**yndrome): caused by the HIV virus. This attacks the immune system so that the body becomes defenceless against infection. Few of those infected with the virus develop AIDS immediately; the great majority are homosexual men (see below*).

AIDS symptoms take 6 weeks to 7 years to develop. They include:

(*a*) weight loss; fever and night sweats; extreme tiredness;

(*b*) a rare kind of pneumonia;

(*c*) skin blotched purple by a rare skin cancer.

All AIDS cases die: no vaccine has been developed yet (but see unit 17.16). The virus can also be passed on by non-sexual methods:

(*a*) the placenta and by breast feeding;

(*b*) any shared puncturing device (ear-piercing, needles for drug abuse);

(*c*) toothbrushes, razors (any chance of bleeding). Saliva, tears, crockery and towels cannot pass the virus on.

2 Genital Herpes: caused by a virus related to chicken pox. Causes extremely uncomfortable sores around the genitals. The simplest way to avoid all STD is not to have casual sex.

15.6 Abortion

Abortion is the ending of pregnancy with the death and removal of an embryo. There are two types.

(*a*) *Spontaneous* (a 'miscarriage'): some abnormality of the embryo or the mother results in the embryo's death.

(*b*) *Induced:* the embryo is removed surgically by a suction device. Two of the legally accepted reasons for permitting an abortion are:

(i) the mother's health is at risk in having the baby;

(ii) the baby is likely to be born abnormal, e.g. with Down's syndrome (see unit 17.15). Amniocentesis can warn of this: cells of the embryo present in a little amniotic fluid, withdrawn by syringe, can be grown in culture. Abnormal chromosomes in these cells may indicate Down's syndrome (mongolism).

Abortion is *not* a means of contraception (see unit 15.6). It carries some risk to the mother's health; it can cause severe depressions, and is regarded by some as murder of the unborn.

*Note: Up to December 1985 an estimated 20 000 people were infected in the UK. Only 275 were AIDS cases (245 homosexual or bisexual men; 14 infected through blood transfusion). In the USA there were over 15 000 cases. Bisexuals may be the cause of the rapid spread of AIDS.

16 GROWTH OF CELLS AND POPULATIONS

16.1 Principles of Growth

Growth: irreversible increase in size or mass of an organism.
Processes involved in growth
1 Formation of more protoplasm, especially proteins: cell size increases.
2 Cell division by mitosis (see unit 17.10): maintains small size of cells.
3 Vacuolation – in plants only; absorption of much water, swelling the cell.
4 Differentiation – cells become different for special purposes (see Fig. 16.1).

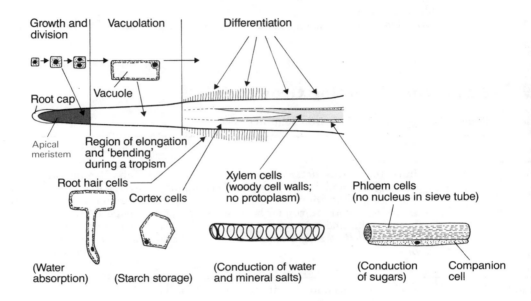

Fig. 16.1 Longitudinal section through a root showing regions of cell division (in red) and subsequent stages in growth

In a similar way, animal cells divide (but do not form vacuoles) and differentiate into cheek cells, muscle cells, neurones and blood cells etc.
All four processes are controlled by hormones (see units 12.10: pituitary, 12.14: auxin).
Growth involves changes of shape as well as size.

A Shape
Plants and animals grow differently to suit their type of nutrition.
1 Green plants grow at their tips giving a branching shape with a large surface area for absorption of nutrients (necessary when anchored), and of sunlight energy.
2 Animals' bodies grow into a compact shape, except for their limbs (needed for food-seeking).

B Size

1 Unicells have a large surface area for their tiny volume, so diffusion of materials meets most of their needs. The same applies to the individual cells of multicellular organisms.

2 Multicells specialize their cells, grouping them into tissues and organs. These meet the same needs of cells more efficiently. Diffusion of food, oxygen and wastes through the body's large volume would be too slow to ensure survival (see unit 9.5).

So special tissues are needed for
(*a*) *absorption and excretion*, e.g. in roots, guts, lungs, leaves, kidneys;
(*b*) *transport*, e.g. blood, xylem, phloem;
(*c*) *support*, e.g. bone, xylem;
(*d*) *coordination*, e.g. hormone producing cells, neurones;
(*e*) *reproduction*, e.g. in genitals, flowers; and as gametes.

Large size gives advantages in nutrition.
Tall plants can starve small ones by shading them.
Large animals can use their greater power to gain more food than small ones; and to respond to predators' attacks.
Small size is useful where food is scarce, and for hiding from predators.

16.2 Factors Affecting Growth

Table 16.1 Some effects of genes and environment on growth

	Plants	*Animals*
Genes	Inherited factors; determine *size*, e.g. tall and dwarf varieties of pea plants – through growth hormones *shape*, e.g. beetroot, runner bean, gooseberry bush, poplar and oak trees *rate of growth*, e.g. pine trees grow faster than oak	Inherited factors; determine *size*, e.g. large and small dogs – through hormones (see unit 12.10) *shape*, e.g. dachshunds and bulldogs *colouring*, e.g. tabby cats and siamese *growth pattern* (see units 16.3 and 21.7) *rate of growth*, e.g. sealions grow faster than humans
Climate	*Light* is essential for nutrition (photosynthesis) and therefore growth *Increased temperature* speeds up metabolism, e.g. respiration and photosynthesis, and thus rate of growth	*Light* is necessary for making vitamin D, needed for bone growth (see unit 4.5) *Increased temperature* speeds up growth and development of ectotherms (unit 10.6) but not endotherms, e.g. mammals
Nutrients	*Mineral salts* of the right kinds (see unit 4.3) and quantity, e.g. from fertilizers, promote growth *Water* and carbon dioxide essential	*Food* of the right kinds and quantity (a balanced diet) promotes growth (see unit 4.6) *Water essential*

Gene expression is affected by:

(*a*) *climate*: stoats change their brown summer coats for white ('ermine') ones for the winter; black markings of dark moths reared in cool conditions are paler (brown) if reared in warm conditions.
(*b*) *nutrients*: both size and intelligence of humans are less when they are continually under-nourished (see unit 4.6); genes cannot express themselves fully.

16.3 Human Growth

Growth from birth is *continuous* (compare insects, see unit 21.7).
Rapid growth occurs at:

(*a*) suckling time;
(*b*) puberty (earlier in girls than boys) – see Figs. 16.2, 16.3.

Given good nourishment, puberty is reached earliest in the Japanese and Chinese, followed by black African races and then Europeans. Under-nourishment delays growth and may reduce expected height and weight if it is continuous. Swedish 17-year-old boys of the 1880s were as much as 15 cm shorter than their 1970 equivalents.

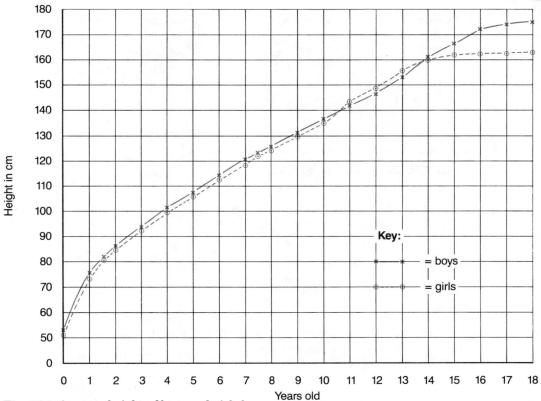

Fig. 16.2 Average height of boys and girls by age

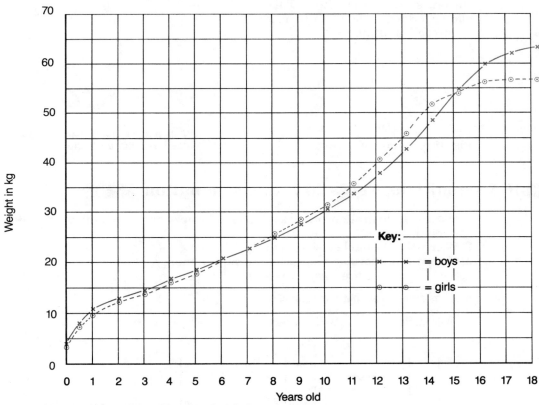

Fig. 16.3 Average weight of boys and girls by age

16.4 Seed Structure and Germination

Seeds are embryo plants enclosed by the testa (seed coat). They develop from the ovule after fertilization (see unit 14.8). When shed they are dry (about 10% water).

The *embryo* consists of a radicle (root), a plumule (shoot) and one or more cotyledons (first seed leaves).

The *testa* bears a hilum (scar where the seed broke off the fruit) and a micropyle (pore for water entry during germination). (See Fig. 16.4.)

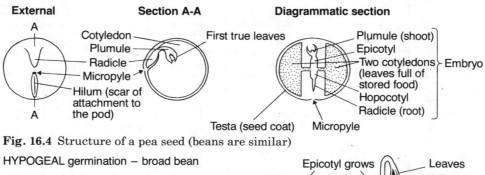

Fig. 16.4 Structure of a pea seed (beans are similar)

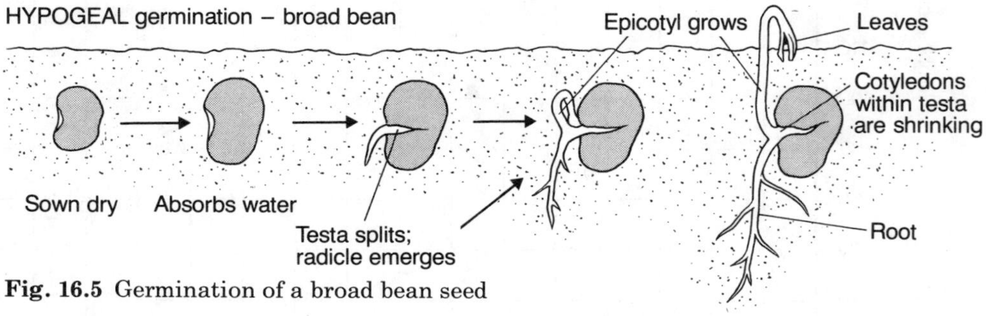

HYPOGEAL germination – broad bean

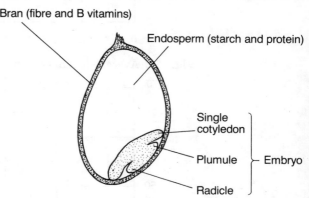

Fig. 16.5 Germination of a broad bean seed

In peas and beans the food for the seedling's growth is stored in the cotyledons. In wheat and maize the food store is the endosperm (outside the embryo). (See Fig. 16.6.)

When barley germinates, enzymes turn its starch into maltose. This 'malt' can be fermented by yeast to make beer (see Fig. 3.12).

Fig. 16.6 A grain of wheat

White bread is made mainly from endosperm. Wholemeal bread uses the whole grain

16.5 Conditions Necessary for Germination

1 **Water:** to hydrate protoplasm, activating enzymes which digest stored food (e.g. starch to sugars).
2 **Warmth:** to enable enzymes to work.
3 **Oxygen:** for aerobic respiration to supply energy for growth.
    Some seeds require *light*, others dark, for germination; for most these do not matter.

All tubes except C are put in a warm place

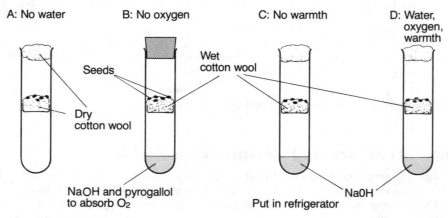

Result: only seeds in D germinate

Fig. 16.7 Experiment to determine the conditions necessary for seed germination

16.6 Growth Measurement and its Difficulties

1 Length or height – all organisms. A crude method: volume would be better.
2 Live mass – terrestrial animals. Difficult for:
(*a*) plants: roots are broken off, or soil remains attached to them;
(*b*) aquatic organisms (how much should one dry them before weighing?).
3 Dry mass – all organisms, but they have to be killed (and dried in an oven at 110°C). Avoids errors of hydration likely in no. 2 above, e.g. Did the animal drink or urinate before it was weighed? Were the plant cells fully turgid on weighing?

It is essential when measuring growth of organisms to
(*a*) have a large number growing (avoids results from a freak individual);
(*b*) control *all* factors affecting growth (including crowding), e.g. food, temperature, light.

16.7 Growth of Populations

The stages in growth of an individual organism are reflected by the changes in population of *cells* within it.

Table 16.2 Stages in growth

	Stage	Cell population
A	Embryo	Rises very rapidly (exponentially)
B	Youth	Rises rapidly but more steadily (new cells formed greatly outnumber those dying)
C	Maturity	Reaches maximum (new cells cancelled out by the same number dying)
D	Senescence (growing old)	Falls slowly (more cells die than are formed to replace them)
E	Death	Falls very rapidly (failure of some part of the body on which all other cells depend)

There are similar stages in growth of a population of *organisms*. An exponential increase is followed by a steadier increase in numbers until a maximum is reached (stationary phase). This maximum is decided by factors in the environment that affect birth rate, death rate (and immigration or emigration). If these effects are severe, the population may fall drastically, even causing extinction (see Fig. 16.8).

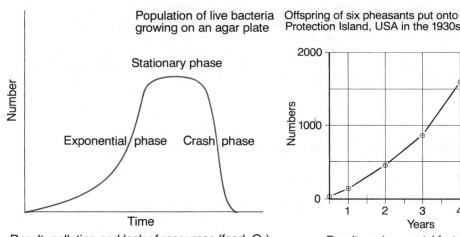

Population of live bacteria growing on an agar plate

Offspring of six pheasants put onto Protection Island, USA in the 1930s

Result: pollution and lack of resources (food, O_2) cause death of the majority of bacteria

Result: environmental factors start to check numbers at year 4

Fig. 16.8 Population of live bacteria growing on an agar plate

Fig. 16.9 Offspring of six pheasants put onto Protection Island, USA in the 1930s

In this artificial (laboratory) example no other organisms, or even the climate, contributed to the crash in numbers of bacteria.

16.8 Human Population

In a natural environment many factors control plant and animal numbers (see Fig. 16.9). Similar factors once controlled the population of Man's ancestors – until Man worked out ways of avoiding their controlling influence (see Table 16.3).

Table 16.3 Man's avoidance of the factors controlling populations in nature

Factors controlling populations in nature	Man's methods of avoiding natural population control
1 Climate	Shelter (homes), clothes, fire
2 Predators	Tools – allowed Man's ancestors to overcome other animals with weapons
3 Food supply	Tools and science – adopting agriculture (which produces much more food than hunting and gathering it, as animals do)
4 Disease	Science and politics – understanding diseases so as to cure and prevent them (medically and by public health laws)

Human population, freed from the severe control imposed by disease (especially bacteria and viruses) and fed by abundant food, has increased rapidly (see Fig. 16.10). This exponential

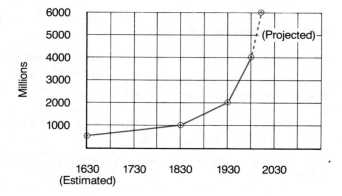

Fig. 16.10 Growth of the human population in the world

population explosion continues. But 'developing' countries are the main ones contributing to it. Since 1975 the 'developed' countries (industrialized, wealthy ones, e.g. in Europe and N. America) have reduced their annual population increases. Some countries in Europe now show no increase at all.

If population growth in developing countries can rapidly be made to slow down (to a near stationary phase) the human race may avoid the fate of the bacteria represented in Fig. 16.8. Modern man must learn to live within the resources of the world without destroying and polluting it. Many 'primitive' races of Man, e.g. the Kung of Botswana, pygmies of African rain forest, bushmen of Australia, Inuit Eskimos and certain Amazonian tribes have all learned to live in balance with their environments. We must re-learn this lesson, to survive. Such issues are dealt with in unit 20.12.

16.9 Population Structure by Age and Sex

Population growth may be worked out by the following equation:

$$\frac{\text{Births} - \text{Deaths} \pm \text{Migration}}{\text{Total population}} \times 100$$

But this figure does not provide governments with enough information to plan for people's future needs, e.g. schools, work, hospitals and care of the aged. The age structure of the population may be shown by **population pyramids** (see Fig. 16.11). **Developing countries** show a characteristic arrow-head shape. They have a high birth rate and high mortality. Emigration can affect the pyramid. **Developed countries** show a more pillar-like shape. They have a low birth rate and low mortality. Immigration can affect the pyramid, but it is now usually restricted by laws.

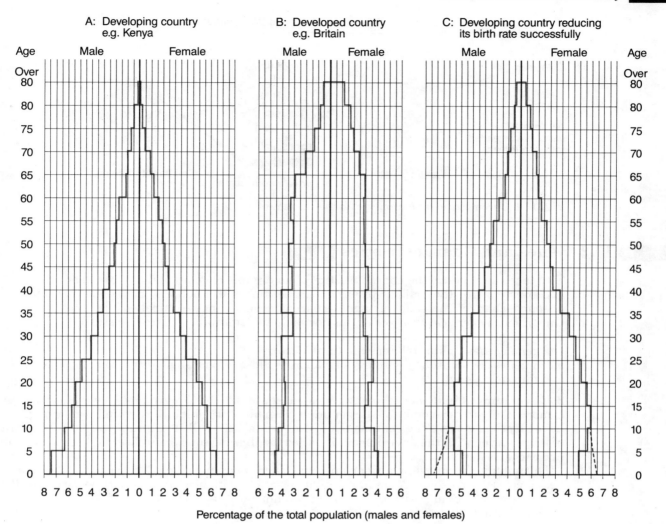

A: Developing country e.g. Kenya

B: Developed country e.g. Britain

C: Developing country reducing its birth rate successfully

Percentage of the total population (males and females)

Fig. 16.11 The general shape of population pyramids for three countries, each at a different stage of development

17 GENES, CHROMOSOMES AND HEREDITY

17.1 The Nucleus, Chromosomes and Genes

The **nucleus** normally contains long threads of DNA (see unit 1.2) which are not visible under the light microscope.

Before cell division each DNA thread coils up, with protein, into a compact 'sausage' called a **chromosome.** When stained, this is visible under the microscope (see Fig. 17.1). Chromosomes are present in **homologous pairs**, both members having the same length (and number of genes). One chromosome of the pair came from the male parent, the other from the female parent, when their gametes fused together to form a zygote, (see Fig. 17.6). Sections of the DNA threads are **genes;** each controls the making of an enzyme (see unit 1.3).

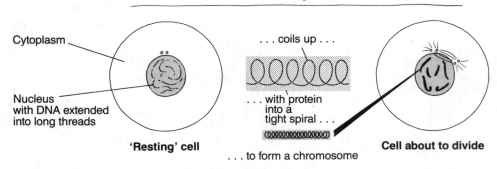

Fig. 17.1 Chromosome formation within the nucleus of a cell

17.2 Genes and Characteristics

Two factors influence the characteristics of an organism: genes and environment.
1 Genes control the making of enzymes (see unit 1.3). Each enzyme controls a particular chemical change. This small change is a part of larger changes controlled by groups of enzymes. One or more chemical changes of this sort help to determine a characteristic.
For example:
one gene causes sickle cell anaemia (unit 17.15);
many genes cooperate to give curly black hair – those for making hair protein, for black colour and for curly growth.

2 Environment influences the way genes act (see unit 17.14). For example, a well-fed youngster is more likely to develop into a larger adult than his starved identical twin (even though both have identical genes). The gene make-up of an organism is called its **genotype.** The interaction of its genotype with the environment is called its **phenotype**, i.e. its observable or measurable **characteristics.**

To simplify the above, consider an unfastened pearl necklace. The pearls are genes; the necklace a chromosome. A similar necklace would be its homologous partner (see Fig. 17.2). Genes at an identical position (locus), on two homologous chromosomes, between them determine a characteristic.

Diagram of chromosomes	a B **Pair 1** ⊂⊃⊂●⊃⊂⊃⊂⊃⊂⊃⊂⊃⊂⊃⊂●⊃⊂⊃ ⊂⊃⊂●⊃⊂⊃⊂⊃⊂⊃⊂⊃⊂⊃⊂●⊃⊂⊃ a b		C_1 **Pair 2** ⊂⊃⊂⊃⊂●⊃⊂⊃⊂⊃⊂⊃ ⊂⊃⊂⊃⊂●⊃⊂⊃⊂⊃⊂⊃ C_2
Genotype	aa Homozygous	Bb Heterozygous	$C_1 C_2$ Heterozygous
Status of these genes	Recessive	B: dominant b. recessive	C_1 and C_2 are co-dominant
Phenotype	a	B	C_1/C_2 (both)

Fig. 17.2 Genetical terms illustrated with reference to two homologous pairs of chromosomes

Dominant genes (shown by capital letters) always express themselves as a characteristic.
Recessive genes (shown by small letters) only express themselves when the partner is also recessive.

Thus genotype **AA** or **Aa** will be expressed as an **A** phenotype and genotype **aa** is the only way of producing the **a** phenotype (see Fig. 17.3). Organisms with two identical genes at a locus (**AA** or **aa** genotypes) are said to be **homozygous;** those with alternative genes at the locus (**Aa**) are called **heterozygous.**

The alternative genes (**A** and **a**) are called **alleles.**

Gene	Phenotype		Genotype	Phenotype		Genotype
Dominant	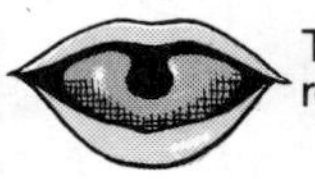 Tongue-roller		RR or Rr	 Free ear lobe		FF or Ff
Recessive	 Non-roller		rr	 Attached ear lobe		ff

Fig. 17.3 Examples of dominant and recessive genes in humans

17.3 Human Blood Groups: Co-dominance

Co-dominance
Certain alleles are **co-dominant**: both alleles express themselves, e.g. in the determination of human blood groups (Table 17.1).

Table 17.1 Genetics of blood groups A, B and O in humans

Gene status	Blood Groups (i.e. phenotypes)	Genotypes
I^A (dominant)	A	$I^A I^A$ or $I^A i$
I^B (dominant)	B	$I^B I^B$ or $I^B i$
i (recessive)	O	ii
I^A and I^B are **co-dominant**	AB	$I^A I^B$

In **blood transfusion** the blood groups of both giver (donor) and the patient receiving blood should ideally be the *same*. This avoids clumping together of the donor's red blood cells inside the patient's veins. Clumping occurs if the patient's antibodies (in the plasma) react with the donor's red blood cells, making their cell membranes sticky.

Clumps not only block capillaries; as the cells in them burst, the released products can cause kidney failure.

Incomplete dominance is not the same as co-dominance: the two alleles express themselves unequally. Examples are shown in Table 17.2.

Table 17.2

Characteristic	Genotypes and their phenotypic effect		
Blood clotting time (see unit 17.9)	HH: normal	Hh: very slightly longer	hh: very long (haemophilia)
Anaemia (see unit 17.15)	AA: none	Aa: very slight	aa: severe (sickle cell anaemia)

Note: The symbols used for genotypes here suggest simple dominance and recessiveness. But the phenotypes in the middle column show this is not strictly true. (At Advanced Level, different symbols are used to allow for this difficulty.)

17.4 Mendel's Experiments

Genetics (the study of heredity) was only put on a firm basis in 1865 thanks to **Gregor Mendel**, an Austrian abbot, who published his research on inheritance in peas.

His materials: *Pisum sativum*, the garden pea. This:
(a) normally *self-pollinates* (and self-fertilizes) when the flower is still unopened. To *cross-pollinate* plants, Mendel had to remove the unripe anthers of strain **A** flowers and dust their stigmas with pollen from ripe stamens of strain **B**. Interference by insects was avoided by enclosing their flowers in muslin bags.
(b) has *strongly contrasting phenotypes*, e.g. pea plants are either tall (180–150 cm) or dwarf (20–45 cm); the seeds are either round or wrinkled.

His methods: As parents (**P₁**, or first parental generation) he chose two contrasting 'pure lines' which 'bred true', i.e. were homozygous. These he mated by cross-pollination. The offspring (**F₁**, or first filial generation) were allowed to self-pollinate. This gave the **F₂**, or second filial generation.

Results from one such experiment:
P₁ Tall × Dwarf
F₁ All Tall

Conclusion 1: factor for Tall is dominant to factor for Dwarf.
F₂ Ratio of 3 Tall: 1 Dwarf

Conclusion 2: factor for Dwarf was not lost (as it seemed to have been in the **F₁**). This suggested that 'factors' were particles of hereditary material which remained unaltered as they were handed on at each generation.

We now know that 'factors' are genes, and that the material of genes is DNA in chromosomes.

Table 17.3 Summary of a Mendelian experiment using modern genetical terms

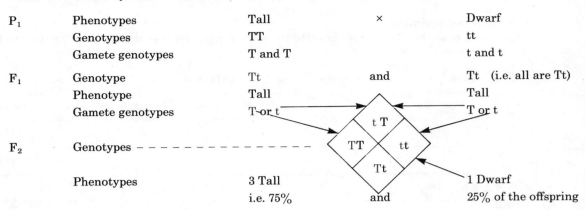

17.5 Hints on Tackling Genetic Problems

When genetics problems are set as questions it is essential that the eight lines of terms relating to the P_1, F_1, and F_2 on the left of Table 17.3 be set out first before the data in the question is inserted in the appropriate places. By reasoning, the rest of the 'form' you have thus created can be filled in. It is vital to remember that gametes are **haploid** (have *one* set of genes) and organisms are **diploid** (have *two* sets of genes). The diamond checkerboard giving the genotypes of offspring is called a Punnett square.

17.6 Back-cross Test

Back-cross test: shows whether an organism of dominant phenotype is homozygous (TT) or heterozygous (Tt). The organism is crossed with a double recessive organism.

Table 17.4

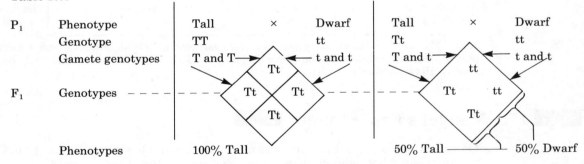

Only heterozygotes can give recessive phenotype (dwarf) offspring. Those that do not (homozygotes) can be used as 'pure line' parents in selective breeding.

17.7 Ratios of Phenotypes

Tables 17.3 and 17.4 state certain ratios of offspring: 75:25 and 50:50. These are only *expected* ratios. The ratios *obtained* in a breeding experiment are rarely identical with those expected. Thus Mendel obtained 787 Tall:277 Dwarf in the F_2 of the experiment explained in Table 17.4, a ratio of 2.84:1. Likewise a coin tossed 1000 times is *expected* to give 500 'heads' and 500 'tails' – but rarely does so. Scientists apply a 'test of significance' to ratios obtained to see whether they are near enough to the expected ratios to be regarded as the same. For example, is 26:24 near enough to 25:25 to be regarded as 50% of each?

Note: You are not expected to know the 'test of significance'. But if you were given a ratio of, say, 505:499 offspring in a question, you must first *explain why* you assume this is a 50:50 ratio before proceeding.

Mendel used *large numbers* of organisms in his experiments to obtain ratios of offspring that were meaningful. Much modern genetical knowledge has come from breeding *Drosophila* (fruit fly) which

(*a*) is easy to culture (in small bottles on banana paste and yeast);
(*b*) produces many offspring (100 per female);
(*c*) has a short development time (10 days from egg to adult).

Investigating the genetics of slow-breeding species (e.g. cows, Man) is less easy. Experiments are lengthy and costly, and in Man's case not allowed. Information must come from herd, family or hospital records.

▦ 17.8 Sex Determination in Mammals ▦

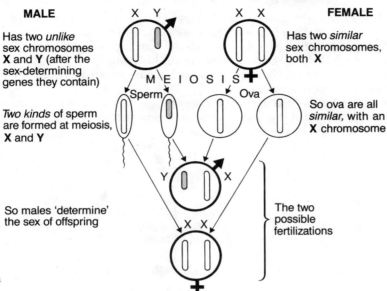

Fig. 17.4
Sex determination

Thus males and females are born in approximately equal numbers.

▦ 17.9 Sex Linkage ▦

Sex linkage: the appearance of certain characteristics in one sex and not the other (in mammals these appear in the male).

The **Y** chromosome, being shorter than the **X**, (see Fig. 17.4) lacks a number of genes present on the longer chromosome. In a male (**XY**) therefore, these genes are present singly and not in pairs, as in the female (**XX**). All these 'single' genes come from the mother (on the **X**) and express themselves, even if recessive. Examples: red/green colour blindness and haemophilia.

In **haemophilia** blood fails to clot; so trivial cuts and tooth extractions can be lethal through bleeding. Nowadays, injections of the clotting factor that they lack (Factor VIII) can help haemophiliacs to lead near-normal lives.

Possible types   H = normal, is dominant to h = haemophiliac

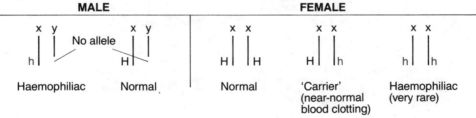

Possible origins

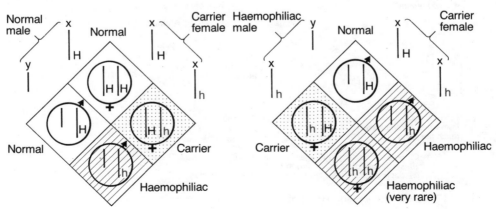

Fig. 17.5
Inheritance of haemophilia

17.10 Mitosis and Meiosis in the Life Cycle

1 Most organisms start from a **zygote** cell containing chromosomes in pairs, i.e. it contains a double set or **diploid** number of chromosomes (2n). One set comes from each parent.

2 The zygote divides by **mitosis** to form new cells, also containing the 2n number, during **growth**.

3 In a multicellular organism these cells **differentiate** (see unit 16.1) into cells as different as neurones and phagocytes. This happens because although all the cells possess identical genes (see unit 17.12), they use different combinations of them according to their location in the body. For example, muscle cells do not use their hair-colour genes.

4 Certain cells in sex organs divide by *meiosis* to become *gametes*. These contain a single set or haploid number of chromosomes (n).

5 At **fertilization**, gametes fuse to form a zygote (n + n → 2n). Meiosis thus ensures that the chromosome number does not double at each new generation (which it would if gametes were 2n, i.e. 2n + 2n → 4n; 4n + 4n → 8n and so on).

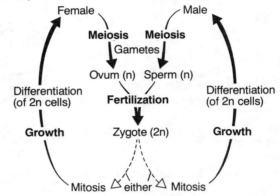

Fig. 17.6 Mitosis and meiosis in a life cycle

Both mitosis and meiosis use similar methods of moving chromosomes; but their *chromosome behaviour* is different.

17.11 How Chromosomes Move Apart at Cell Division

The stages shown in Fig. 17.7 are all part of a continuous process.

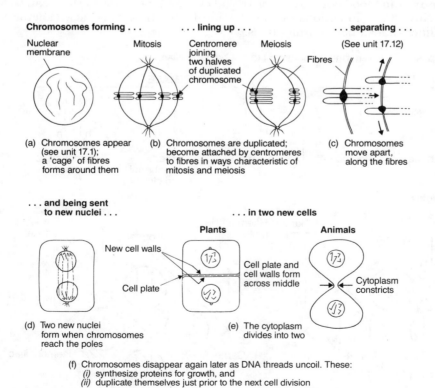

(f) Chromosomes disappear again later as DNA threads uncoil. These:
  (i) synthesize proteins for growth, and
  (ii) duplicate themselves just prior to the next cell division

Fig. 17.7 How chromosomes separate at cell division

17.12 Mitosis and Meiosis compared

Table 17.5 Summary comparison of mitosis and meiosis

	Mitosis	*Meiosis*
Number of cell divisions	1	2
Resulting cells are	Diploid, identical	Haploid, not identical
Purpose	Growth, replacement (e.g. of skin, blood cells) Asexual reproduction	Gamete formation
Occurrence in	Growth areas (unit 16.1) Replacement tissues, e.g. skin, bone marrow Where runners, tubers form	Gonads, e.g. testes, ovaries; anthers, ovules (units 15.1, 14.5)

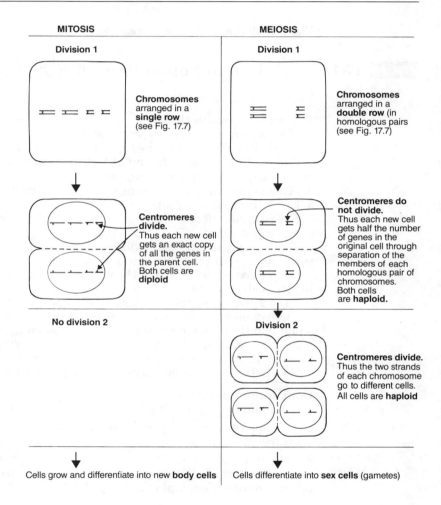

17.13 Meiosis Shuffles Genes

In addition to halving the chromosome number, meiosis shuffles genes at the stage when chromosomes 'line up' (see Fig. 17.7):

(*a*) **by varying the way they line up:**

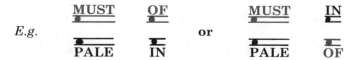

In a human cell with 46 chromosomes (and not 4) this alone makes 2^{22} (over 4 million) different varieties of gamete.

(*b*) **by exchanging material between pairs of chromosomes** ('crossing over')

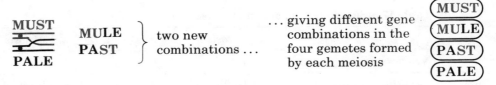

... giving different gene combinations in the four gemetes formed by each meiosis

Other combinations are also possible, e.g. MALE, PUST; MUSE, PALT. In a human cell the number of possible combinations is enormous since most of the 23 pairs of chromosomes have hundreds of genes.

Key: MUST and PALE } homologous chromosomes   M and P } alleles

Fig. 17.8 How meiosis shuffles genes to give an enormous variety of gametes

17.14 Variation in Populations

Variety arises from

1 Sexual reproduction:
Male and female have *different genes*.
Meiosis shuffles their genes to make *gametes different* (unit 17.13).
Each pair of gametes fusing at fertilization gives a different *combination* of genes from any other pair. So offspring are unique.

2 Mutation: inheritable changes that are 'new' (see unit 17.15).

3 Environmental effects
Food supply affects size, e.g. poor soil grows small crops; malnutrition causes obesity or deficiency diseases.
Temperature affects metabolism, e.g. warmth speeds growth; affects colour, e.g. dark moths turn out paler if reared in warm conditions.
Overcrowding affects size through competition for food and water; affects behaviour, e.g. locusts will migrate.

1 and **2** give inheritable variation. This can affect evolution (see unit 18.3). **3** is not inheritable.
Environmental effects can best be demonstrated using a clone (see unit 14.2). Cuttings of equal length from a *Tradescantia* plant, rooted in sand, can be grown in different mineral salt solutions (see unit 5.8). Any differences in growth must be due to differences in the mineral salts present, i.e. environment, since the genes in each plant cutting are identical.

Two patterns of variation are seen in populations:

Continuous variation	**Discontinuous variation**
E.g. human height, intelligence, fingerprints.	*E.g.* height of pea plants; human blood groups.
1 A complete *range* of types, e.g. from giants to dwarfs in humans.	**1** Sharply *contrasting* types, e.g. tall and dwarf pea plants.
2 Phenotype controlled by	**2** Phenotype controlled by
(*a*) *many pairs* of alleles;	(*a*) a *single pair* (or a few) alleles;
(*b*) environment (may play a major part).	(*b*) environment (plays little part).

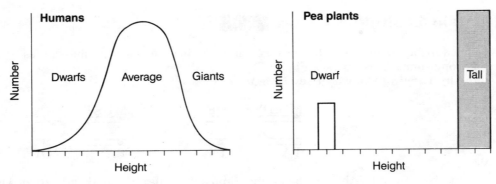

Fig. 17.9 Continuous and discontinuous variation in populations

17.15 Mutation

Mutation: an inheritable change in a cell. The nature of the DNA, or its quantity, alters.

Cause: cosmic rays, ultra-violet rays, X-rays, radioactive emissions and certain chemicals, e.g. mustard gas, are mutagenic agents (cause mutations). These (and other causes) have effects on

(*a*) **Genes:** a minute part of DNA once altered, may produce altered proteins in the cell (see unit 1.3). These may be useless, e.g. *haemophilia* (no Factor VIII for clotting), useful, e.g. *melanism* (black pigment in peppered moths), or a bit of both, e.g. *sickle cell anaemia* (haemoglobin in red blood cells changed) – see unit 18.2.

(*b*) **Single chromosomes:** through accidents at the separation of chromosomes in meiosis, extra chromosomes reach gametes, e.g. older mothers have a higher chance of producing *mongol* (Down's syndrome) babies by this means. Mongols have 2n + 1 chromosomes, i.e. 47. They are moonfaced, physically and mentally retarded and very affectionate.

(*c*) **Whole sets of chromosomes:** meiosis may go completely wrong and produce diploid (2n) gametes. If these are used in fertilization

2n + n (normal gamete) → 3n (triploid chromosome number).

Many apple varieties are 3n.

Cells with more than 2 *sets* of chromosomes (polyploids) are larger than diploid (2n) ones. So Man encourages and selects for these mutations to give bigger crops, e.g. wheat is 6n.

Man also 'mutates' cells by genetic engineering (unit 17.16).

Mutations occur constantly in populations. New strains of viruses, e.g. 'flu, catch human defences unprepared. New strains of bacteria, e.g. syphilis, are resistant to antibiotics. Mutant pest insects, e.g. mosquito, are surviving insecticides; and mutant rats in the UK have become Warfarin (poison) resistant.

17.16 Genetic Engineering

Genetic engineering is Man's transfer of useful genes from one organism into another. Two main methods:

1 (*a*) **Bacteria** can be given selected genes using plasmids (see Fig. 17.10). Inside the bacterium the genes function normally. They are reproduced when bacteria divide. The wanted products (e.g. insulin, growth hormone, vitamins or enzymes) can be extracted in large quantities from the fermenter (in which the bacteria reproduce rapidly – see unit 3.12).

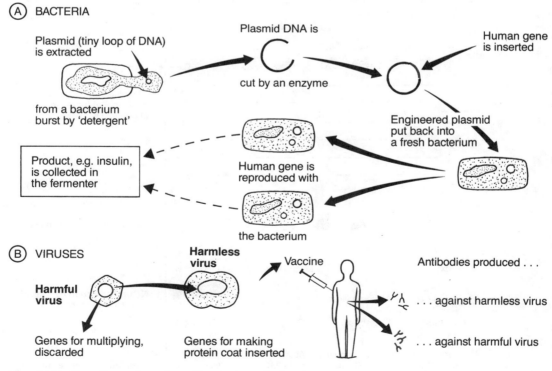

Fig. 17.10 Genetic engineering of bacteria and viruses

(*b*) **Viruses:** Genes for the protein coat of a harmful virus, e.g. AIDS, can be put into a harmless virus, e.g. cowpox. If this 'engineered' virus is used as a vaccine, the body produces antibodies against the harmful virus. Such ideas are on trial.

2 Fusing two kinds of cells together. Lymphocytes producing particular antibodies (e.g. against measles) can be fused with cancer cells (which divide rapidly). These 'engineered' cells both rapidly divide and produce antibodies in very large quantities. The antibodies can be extracted to make vaccines.

Plant cells can also be fused – after removing the cell wall. Attempts are being made to produce a cereal plant with the ability to fix nitrogen.

18 EVOLUTION

18.1 Selection of the 'Best' from a Variety

Organic evolution is the change in a population of a species (over a large number of generations) that results in the formation of a new species. The change is brought about by selection of only the 'best' from the variety of types present in the population.

Variety in a population arises from:

(a) *new combinations of genes* at fertilization (mother and father are different);
(b) *meiosis* (it shuffles the genes to make all gametes different – see unit 17.13);
(c) *mutation* (which includes the origin of 'new genes' – see unit 17.15);
(d) *environmental influences* (other than mutagenic agents), e.g. poor soil grows stunted plants. Note: (a) and (b) are *rearrangements* of existing genes; (c) is the creating of *new* genes; but (d) does not change genetic material at all.

Natural selection occurs when the *environment* permits only those best adapted to it to survive. The survivors are the 'fittest and best' for the particular conditions of the environment at that time (see Fig. 18.1). If the conditions change, new kinds of survivors appear since the previous survivors may no longer be the fittest (see units 18.2A and E).

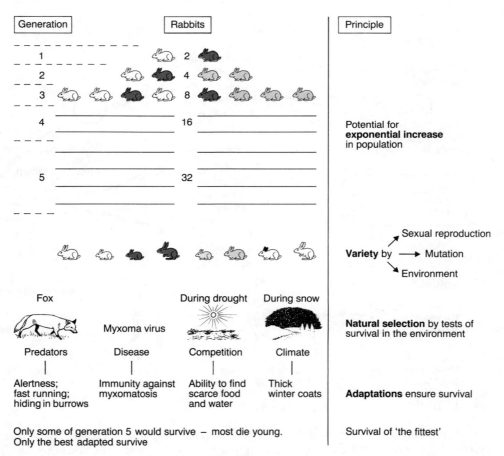

Fig. 18.1 Natural selection results in survival of the fittest

An **adaptation** is a solution to a biological problem. Adaptations may be *structures*, e.g. wings for flying, *chemicals*, e.g. antibodies against diseases, *features of the life history*, e.g. high reproductive rate, and even *behaviour*, e.g. phototaxis in fly adults (see unit 12.13). The majority of adaptations are *inherited*.

Artificial selection occurs when *Man* selects, for his own purposes, certain varieties of organism. These are frequently 'unfit' for survival in the wild, being suited only for the special conditions he puts them in, e.g. farms, greenhouses, gardens and homes (see unit 18.6).

18.2 Examples of Natural Selection

A Industrial melanism in the peppered moth *Biston betularia*
Selection by: predatory birds, e.g. thrushes
Adaptation: camouflage
Before 1840 the light-coloured (peppered) moth survived by camouflage on lichen-covered tree trunks in unpolluted woods (see Fig. 18.2). Dark mutant moths did not survive predation – they were easily seen.

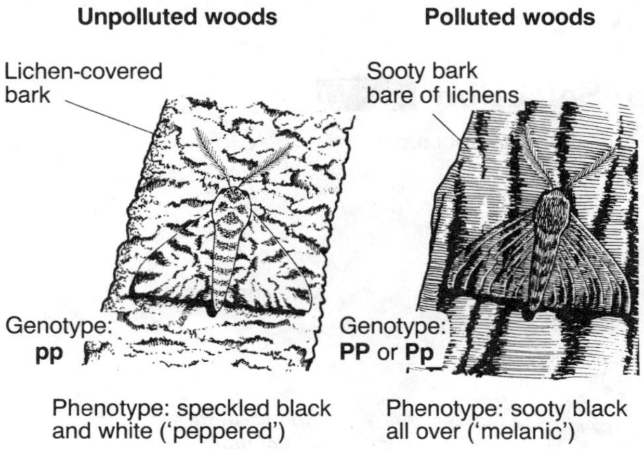

Unpolluted woods   **Polluted woods**

Lichen-covered   Sooty bark
bark   bare of lichens

Genotype: Genotype:
pp   **PP** or **Pp**

Phenotype: speckled black   Phenotype: sooty black
and white ('peppered')   all over ('melanic')

Fig. 18.2 The two forms of the peppered moth, camouflaged

When industrial pollution killed lichens and blackened bark the dark mutant moths (dominant gene, **P**) became camouflaged and survived. The light-coloured moths (**pp**) became obvious to predators. By 1900 few light moths survived in industrial areas and 98% were **PP** or **Pp**. As pollution is being reduced, these mainly dark populations are becoming peppered again; the dark moths are being predated in the cleaner woods and light ones are now surviving better.

B Sickle cell anaemia and human survival
Selection by: disease
Adaptation: type of haemoglobin
Normal red blood cells are plate-like (see unit 8.2) and carry O_2 well. If haemoglobin is abnormal the red blood cells change to half-moon or sickle shapes in capillaries; and O_2 is not supplied properly to cells. A mutant gene is responsible for this fatal condition.

Table 18.1

	Normal person	*Sickle cell trait*	*Sickle cell anaemia*
Haemoglobin:	Normal	Nearly normal	Abnormal
Diseases:	Not anaemic but can die of malaria	Very mildly anaemic and malaria resistant	Severely anaemic (dies)
Genotype	*HH*	*Hh*	*hh*

Thus the homozygote *HH* is at a slight advantage in countries without malaria, and the heterozygote *Hh* is at an advantage in malarial areas.
C Foxes at different latitudes
Selection by: climate – temperature
Adaptations: fur; ears
Arctic foxes: are insulated so well that they tolerate sitting in snow at −40°C. Ears small to retain heat (and not freeze).
Red foxes: fur can keep bodies warm in European winters. Ears medium size.
Kit foxes: fur not very thick – keep cool from desert heat in burrows. Ears large – radiate heat (and hear well at night).

D Red deer and sexual selection
Selection by: sexual competition
Adaptations: antlers and powerful bodies
Males (stags) push each other, antlers against antlers, to fight for a group of females (hinds) to mate with. The strongest gain most mates, so passing on their genes; the weakest do not mate.
E Shortage of food for deer on the Kaibab Plateau, Arizona
Selection by: competitors for the food
Adaptation: ability to get enough food (including standing up on hind legs to browse trees)
    In 1900, about 4000 well fed deer roamed the plateau. Predators, e.g. wolves and pumas, kept deer numbers steady. In 1907, the majority of the predators were killed by Man. Unchecked, the deer population rose to over 100 000 in 20 years. Severe competition for food on an over-grazed plateau resulted in a rapid fall in population.
Note: Before 1907, deer able to outwit predators had been selected. The new conditions of 1907–1927, however, selected for deer able to compete successfully for the scarce food.
F Pesticides and medicines (produced by Man) are also agents of 'natural' selection (see unit 17.16). Certain mosquitoes in South America are now adapted to surviving five major insecticides, and many kinds of disease bacteria have become resistant to certain antibiotics.
    Note how mutation and new combinations of genes in all the examples A–F provide the basis for adaptation (and thus survival).

18.3 Evolution by Natural Selection

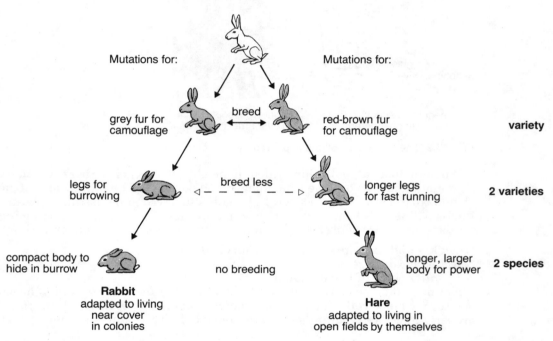

Fig. 18.3 Accumulation of different mutations in two populations over a large number of generations leads to the evolution of new species.

Tests of natural selection	1 The environment provides tests of survival. These include climate, soil type, predation, disease and competition (see unit 18.2).
Variety	2 Of the variety of types in a population, only certain types will survive these tests.
Survival by adaptation	3 These 'selected' types survive because they are adapted to the environmental conditions. They breed, passing on their genes for these adaptations.
Mutation (*'new' genes*)	4 Any mutation which helps an individual organism to survive also gives it a better chance to breed.
Mutants survive and multiply	5 By breeding, these new genes are spread in the population. 6 The numbers of mutant organisms (those having the genes helping survival) rises. So the population as a whole changes its characteristics, i.e. it evolves (see unit 18.2A).
Distinct varieties ...	7 In two *different* environments, the mutations that help survival are also likely to be different (see Fig. 18.3). So two main varieties

arise as they gather their own, different, mutations and become more and more different.

... inter-breed less

8 As the two populations become increasingly different, less and less successful interbreeding between the two marked varieties occurs. (Any hybrids (crosses) tend not to be as well adapted to the two environments as the two specialized varieties, so they do not survive.)

Barriers to inter-breeding

9 Any breeding barrier (e.g. the Pyrenees mountains between France and Spain) that prevents intermixing of genes between two populations speeds up the process of varieties becoming increasingly different.

New species

10 After many generations, the differences between the varieties become so great that they cannot interbreed. Two new species have been formed.

18.4 Charles Darwin (1809–82)

As naturalist on *HMS Beagle* (1831–36), Darwin collected much evidence around the world of 'modification of species by descent' (see unit 18.5).

In 1839, Darwin read *An essay on the principle of population* by the Reverend T. Malthus. This suggested that

(*a*) the human population could increase exponentially (e.g. 2, 4, 8, 16, 32);
(*b*) but the resources, e.g. food, for it could only increase arithmetically (e.g. 2, 3, 4, 5, 6).

As the population's needs could not be met by the resources, there would be a struggle to survive. Famine, disease and war would control the population, unless 'moral restraint', e.g. late marriage, were practised (whose modern equivalent could be contraception).

This idea of struggling to exist provided Darwin with the idea that in similar circumstances in nature the fittest organisms would survive. Darwin's ideas on the 'Origin of Species' are summarized below:

Observation 1 All organisms could, theoretically, increase in numbers exponentially, i.e. **organisms produce more offspring than could possibly survive.**

Observation 2 Populations of organisms, in fact, remain reasonably constant.

Deduction 1 Organisms must have to **struggle for survival** against factors that check their increase in numbers.

Observation 3 In any population there is a variety of types. Much of this variation is inherited by future generations.

Deduction 2 Those best adapted to their environment will survive, i.e. **survival of the fittest**.

Observation 4 Some species have more than one distinct variety.

Observation 5 Anything hindering interbreeding between two varieties will tend to accentuate their differences because each variety will accumulate mutations, many of which will be different between the two varieties.

Deduction 3 New species arise when **divergence of the two varieties** is great enough to prevent interbreeding between them.

Darwin recognized that evolution is a *branching* process. Modern types of ape, e.g. gorillas, did *not* give rise to Man, but both Man and gorilla are likely to have had common ancestors in the distant past (see Fig. 18.4).

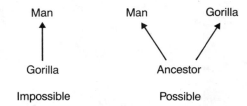

Fig. 18.4 Origins of Man

18.5 Evidence for Evolution

1 **Fossils:** organisms from the distant past whose hard parts (and sometimes impressions of soft parts) are preserved in sedimentary rocks. They are rarely found since dead organisms usually disintegrate by decaying. The oldest fossils known are blue-green algae (similar to bacteria) from 1600 million years ago. The fossil record (see Fig. 18.5) clearly indicates that:

(*a*) the variety of life today probably did *not* arise all at once (as suggested in Genesis in the Bible);
(*b*) first life was aquatic. Terrestrial groups came later (fern-like plants, insects and amphibia arose about 400 million years ago);

(*c*) groups with improved adaptations increase in importance while those less efficient decline (or even become extinct). Improvements in methods of reproduction illustrate this. Thus ferns, dependent on water for fertilization (see unit 21.2), gave way to conifers whose gametes are enclosed in a pollen tube and thus need no water to swim in. Similarly, amphibia gave way to reptiles as the dominant land group. Fossils of pollinating insects and of insect-pollinated flowers are of approximately the same age. These 'coincidences' can be explained in evolutionary terms.

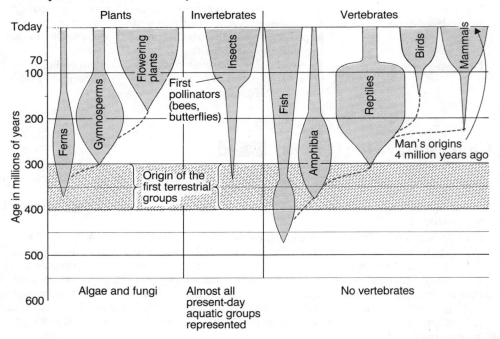

Fig. 18.5 Selected features of the fossil record. The width of each area above represents increase or decrease in abundance of the group

2 The **pentadactyl limb** of vertebrates has a common structure: five 'fingers', wrist bones and one upper and two lower long bones. Yet the purposes to which they are put are very different in bats, Man and whales. Their limbs are thought by evolutionists to have been 'modified by descent', i.e. they are *homologous:* same origin, but different purposes. Wings of bats and butterflies have the same purpose but different structural origins and are *analogous* (see Fig. 18.6).

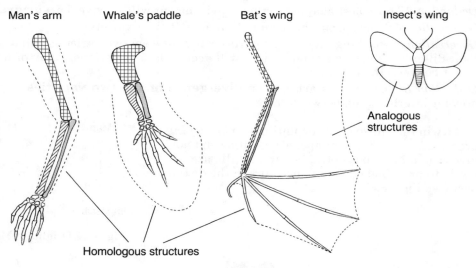

Fig. 18.6 Pentadactyl limbs: homology and analogy

18.6 Artificial selection

Selective breeding by Man has produced:

(*a*) **Plants:**
by *cross-breeding* strains with desirable characteristics
by increasing the *mutation* rate in stamens, using radioactive materials

by vegetative propagation of new strains thus obtained
by *genetically engineering* microorganisms

(i) climate-adapted crops, e.g. lettuce for cool and hot conditions
(ii) disease-resistant crops, e.g. strawberries resistant to viruses
(iii) high-yielding crops, e.g. rice plants that do not blow over in wind, so spoiling the rice grains (they also grow fast enough to allow two or three crops per year instead of one); wheat and apples (see unit 17.15)
(iv) nutritious crops, e.g. maize strains containing *all* the essential amino-acids, so helping to fight kwashiorkor (see unit 4.6)
(v) purpose-selected organisms, e.g. yeasts for brewing, for baking and for SCP; bacteria 'engineered' to make insulin (see units 3.12 and 17.16)
(vi) attractive plants, e.g. roses, chrysanthemums and appetizingly coloured fruits

(b) **Animals:**
by cross-breeding strains and breeding from interesting mutants

(i) dogs as different as bulldogs, dachshunds, St Bernards and Afghan hounds
(ii) horses such as Shetlands, shires, race-horses and mules
(iii) cattle for milk (Jerseys), for beef (Herefords) and for resistance to trypanosome diseases in Africa (Zebu × Brahmin)

and by improving not only genetic stock but also the means of feeding and care

(iv) hens laying more eggs per year (180 in 1920; 280 in 1980)
(v) cows producing more milk per year (2500 kg in 1920; and 4500 kg in 1980)

19 ECOLOGY

19.1 The Biosphere – (its Limits and Organization)

Ecology is the scientific study of organisms in relation to their environment.

Environment: the influences acting upon organisms. Two kinds:

(a) **biotic:** other organisms such as predators, competitors, parasites.
(b) **abiotic:** non-living influences, such as climate, soil structure and water currents.

Habitat: the particular type of locality in an environment in which an organism lives, e.g. among weeds in a pond (stickleback) or among rotting leaves or wood in woodland (woodlouse).

Usually every species of organism exists as a **population** in its environment and not just as a single individual. Together, all the populations of all the species interact to form a **community** within their ecosystem. The role each species plays in the community is its **niche,** e.g. an earthworm's niche is to affect soil fertility by its activities (see unit 19.12) and provide food for shrews, moles and some birds.

Ecosystem: any area in which organisms interact both with each other *and* with their abiotic environment, to form a self-sustaining unit (see Fig. 19.1).

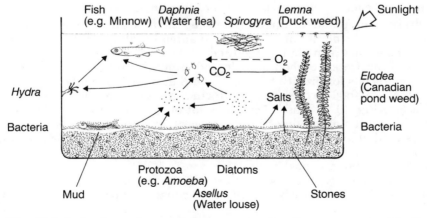

Fig. 19.1 A simple pond ecosystem in an aquarium (arrows represent feeding)

Examples of ecosystems: ponds, jungle, ocean or even a puddle. Ecosystems are not actually distinct, they interact with others. Thus dragonfly nymphs in a pond emerge as flying predators which catch their insect prey over both pond and meadow, so linking both these ecosystems. Even ecosystems in the UK and Africa are linked – by the same swallows feeding on insects in both areas according to the time of year.

Biosphere: the earth's surface that harbours life – a very thin layer of soil and the oceans, lakes, rivers and air (see Fig. 19.2). The biosphere is the sum of all the world's ecosystems and is isolated from any others that may exist in space. However, other celestial bodies influence it:
(*a*) life depends on solar energy (from the sun);
(*b*) other radiations from various sources cause mutations;
(*c*) gravitational fields of sun and moon cause tides;
(*d*) at least 200 tonnes of cosmic dust arrive on earth daily.

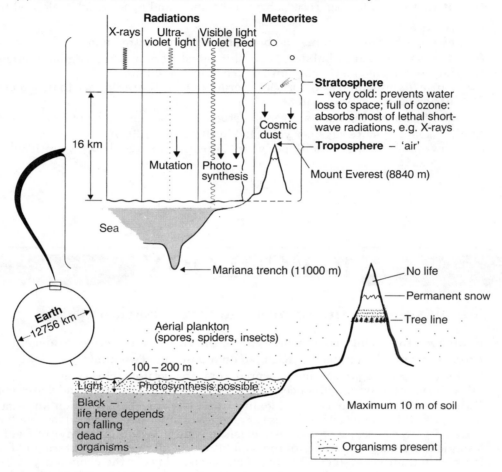

Fig.19.2 The biosphere in relation to the earth

19.2 Food Chains, Food Webs and Food Cycles

Food chains, webs and cycles are units composing an ecosystem.

1 Food chain: a minimum of three organisms, the first always a green plant, the second an animal feeding on the plant, and the third an animal feeding on the second (see Fig. 19.3).

All life depends on green plants (**producers**). They alone can trap sunlight *energy* and make organic *food* from water, carbon dioxide and mineral salts. Animals (**consumers**) get their energy and materials for growth from the food that producers make.

Every transfer of food up the chain results in a great loss in mass (**biomass**) – anything up to 90 per cent (see Fig. 19.4). This is because a lot of food consumed by animals is lost owing to respiration, excretion and indigestibility (faeces), and never reaches the next member of the chain. Thus food chains can be expressed quantitatively as **pyramids of numbers** or, more usefully to farmers and game-wardens, as **pyramids of biomass** (see Fig. 19.3). Such considerations of quantity (of organisms) explain why:
(*a*) the number of species in a food chain rarely exceeds five;
(*b*) the biomass of each species is limited by the capacity of producers (green plants) to produce food;
(*c*) an omnivore gains more food by being a vegetarian than by eating meat. Man would do better to eat grain rather than eat cattle fed on the grain (assuming first class protein needs are met – see unit 4.6).

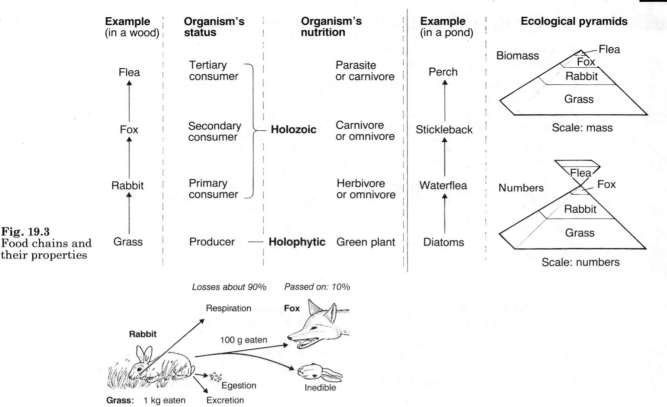

Fig. 19.3 Food chains and their properties

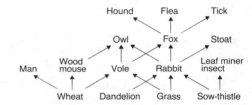

Fig. 19.4 Loss in biomass at each step in a food chain

2 Food web: a number of interlinked food chains. In an ecosystem that includes foxes and rabbits, the diet of consumers is usually more varied than a food chain suggests. Foxes eat beetles, voles, chickens and pheasants as well as rabbits; and rabbits eat a great variety of green plants (see Fig. 19.5).

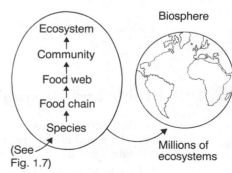

Fig. 19.5 Food web: a number of interrelated food chains

3 Food cycles: food chains with **decomposers** added –
(*a*) *Detritivores:* animals which eat dead and decaying organisms, e.g. water-louse, woodlouse, earthworms, springtails.
(*b*) *Saprophytes:* fungi decay plant materials; bacteria decay protein especially. Thus, dead organisms and excreta (organic matter) are turned into mineral salts and CO_2 (inorganic matter) – which producers need for food but could not otherwise obtain.

4 Energy chain: the passage of energy from the sun along a food chain and on to decomposers. Energy is *not* cycled (Fig. 19.6). It is progressively lost along the chain, e.g. as heat from respiration.

The units making up the biosphere may be summarized as in Fig. 19.7.

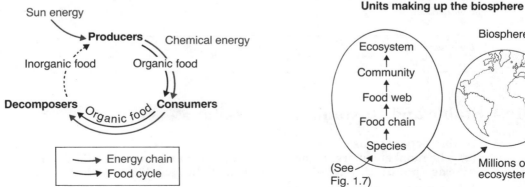

Fig. 19.6 The energy chain in a food cycle

Fig. 19.7 Units making up the biosphere

19.3 Feeding Relationships Between Species

1 Predation: a *predator* is usually larger than its *prey*, an organism it kills for food, e.g. fox kills rabbit; heron kills perch. *Note:* Both organisms are animals, never plants.

2 Parasitism: a *parasite* is an organism living on or in another organism called its *host*, from which it derives its food usually without killing it.
Examples: mosquito (see unit 21.12), green-fly (ecto-parasites); tapeworm (see unit 21.15), leaf-miner caterpillar (endo-parasites).

3 Symbiosis (mutualism): a *symbiont* and its partner (also a *symbiont*) live very closely together, mutually helping each other. Examples:

Symbiont	Nitrogen-fixing bacteria	Mycorrhiza fungi (around roots)
Exchange	Nitrates Sugars ↑	Phosphate ions Sugars ↑ absorbed from soil
Symbiont	↓ Legumes, e.g. clover	↓ Trees, e.g. oak, pine

(See Fig. 20.2)   See (Fig. 19.11)

4 Competition: occurs between two organisms (*competitors*), both attempting to obtain a commodity which is in *short supply* in the environment. Examples:

Commodity	*Competitors* (may be members of *own* species; or of *different* species)	
Light	Waterlilies out-shade diatoms	Oaks out-shade hawthorns
Food	Stickleback and minnow (for water-fleas)	Squirrels and wood pigeons (for acorns)
	Water-fleas (for diatoms)	Tits (for caterpillars)
Nesting sites	Sticklebacks for nest materials (unit 21.16)	Tits for tree-holes
Mates	Frogs	Blackbirds

See also unit 18.2D – mates, and unit 18.2E – food.
   The planting density for crops is determined so as to minimize competition between individual plants. Weeds are successful competitors of crops.

5 Commensalism: a loose relationship between two organisms in which the *commensal* (smaller) benefits by feeding on scraps of food wasted by the *host* (larger) – who is neither harmed nor helped, e.g. sparrows feed on Man's discarded bread. Commensals have alternative food, so the relationship is not obligatory.

Table 19.1 Summary comparison of feeding relationships between organisms
(+ = benefits, − = harmed, ○ = unaffected; A > B means A is larger than B)

	Organisms A	B	*Size relationship*
Predation	Predator +	Prey −	A > B
Parasitism	Parasite +	Host −	A < B
Symbiosis	Symbiont +	Symbiont +	Any
Competition	Competitor −	Competitor −	Any
Commensalism	Commensal +	Host ○	A < B

19.4 Stable and Unstable Ecosystems

Stable ecosystems
(*a*) The numbers of organisms rise and fall (according to season) but a **more or less constant average population level** is maintained. As light and temperature increase from early spring:
1 the numbers of producers increase until early summer when herbivores help to check the increase (see Fig. 19.8);

2 the numbers of herbivores increase until mid-summer when carnivores begin to check their numbers;

3 carnivores only begin to increase in numbers when their prey is sufficiently abundant.

As light and temperature fall to autumn levels, numbers of organisms begin to fall towards early spring levels. Less producers means less food for herbivores; less herbivores means less food for carnivores. Deaths by freezing and starvation, and the dying down of leaves before perennation, provide decomposers with food – which is converted to mineral salts.

(*b*) On land, a **dominant plant species** emerges, e.g. oak, which cannot be out-competed. At this stage a *climax community* exists.

Ponds are always potentially unstable, through silting up.

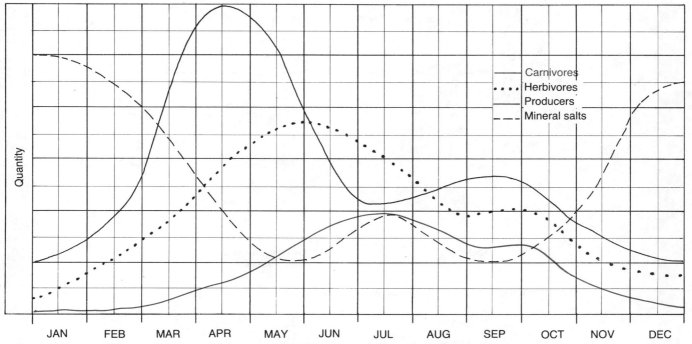

Fig. 19.8 Rise and fall of quantities of food in a pond ecosystem during a year

Unstable ecosystem
Large changes in numbers of most species owing to a changing environment. Examples:
(*a*) *Succession* – one dominant group of species out-competes another group, and changes its environment.

> E.g.   **pond**   ^{silting}   **marsh**   ^{drying}   **oakwood**
> (with *Elodea, Spirogyra, Daphnia*) ⟶ (with bullrush, *Iris*) ⟶ (with oaks)

(*b*) *Pollution or disease* – may disrupt the food web by causing the death of important organisms, e.g. sooty smoke from industry kills lichens (see unit 18.2A); or Dutch-elm disease kills elms; or sewage causes eutrophication (see Table 20.3).

Some causes of instability
(*a*) *Removing a member of the food chain* affects the whole chain, e.g. shooting foxes kills their fleas; allows rabbits to increase; and grass will become over-grazed.
     Thus an effect on one organism has an effect on all.
(*b*) *Simple ecosystems* (i.e. with few food chains), e.g. the terrestrial Arctic or Man's monocultures on farms, can most easily be upset by removal or addition of species. Complex ecosystems are more stable, e.g. tropical forest, owing to a great variety of alternative foods.
(*c*) *Erosion of soil*, e.g. by deforestation on sloping ground, removes the basis for plant growth and thus destroys whole ecosystems.
(*d*) *Pollution of air and land* often affects water by drainage, e.g. acid rain, insecticides (see unit 20.6).

▮ 19.5 Pond Ecosystem ▮

Although Fig. 19.9 shows some features of ponds, you are unlikely to see them all. Fig. 19.10 suggests how a class study may be carried out; but you may not have a flat-bottomed boat or pier to work from.

Much useful information can be got by using a dipping net (wide mesh) from the bank and establishing an aquarium with mud and plants in week 1, and representative animals (not fish) in week 2. Fish, e.g. stickleback, can be kept separately in their own aquarium and fed separately.

There are suggestions for experiments of your own based on ponds in unit 22.9.

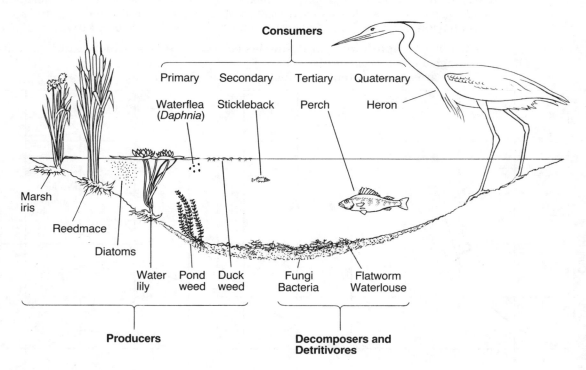

Fig. 19.9 Diagram of a simple pond ecosystem

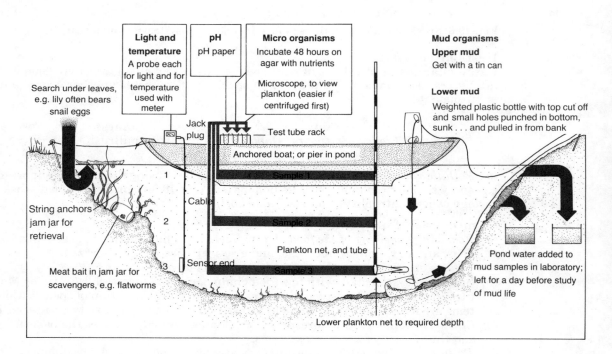

Fig. 19.10 Some methods for studying the ecology of a pond

19.6 Woodland Ecosystem

Figure 19.11 gives some idea of the general structure of a wood, but the one you study will have differences. A wood is really two related ecosystems: woodland above and soil below. Figures 19.12–19.14 suggest how you may set about a class study of a wood. In unit 22.9 are suggestions for experiments of your own.

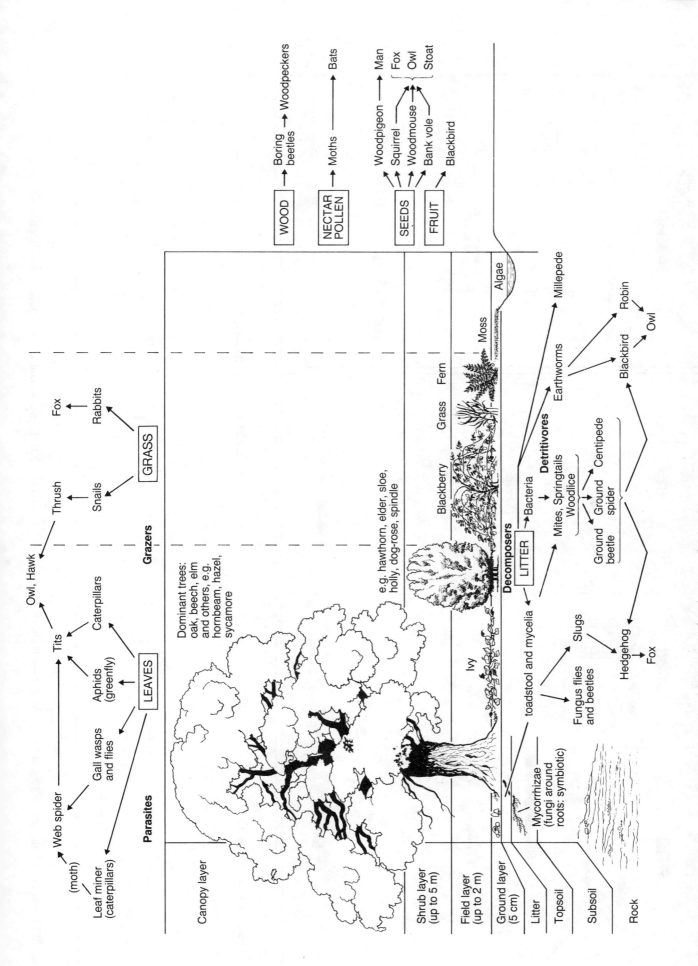

Fig. 19.11 Structure of a deciduous woodland ecosystem showing some food webs

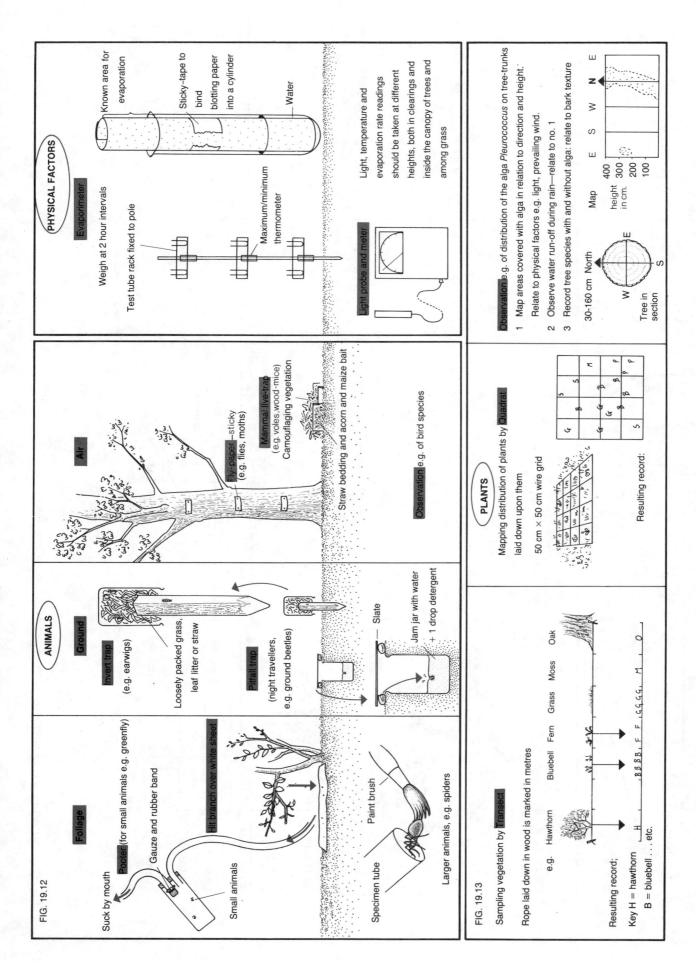

Fig. 19.12 Some methods for studying the ecology of a wood above ground

Fig. 19.13 Some methods for mapping vegetation in a wood

19.7 Soil Ecosystem

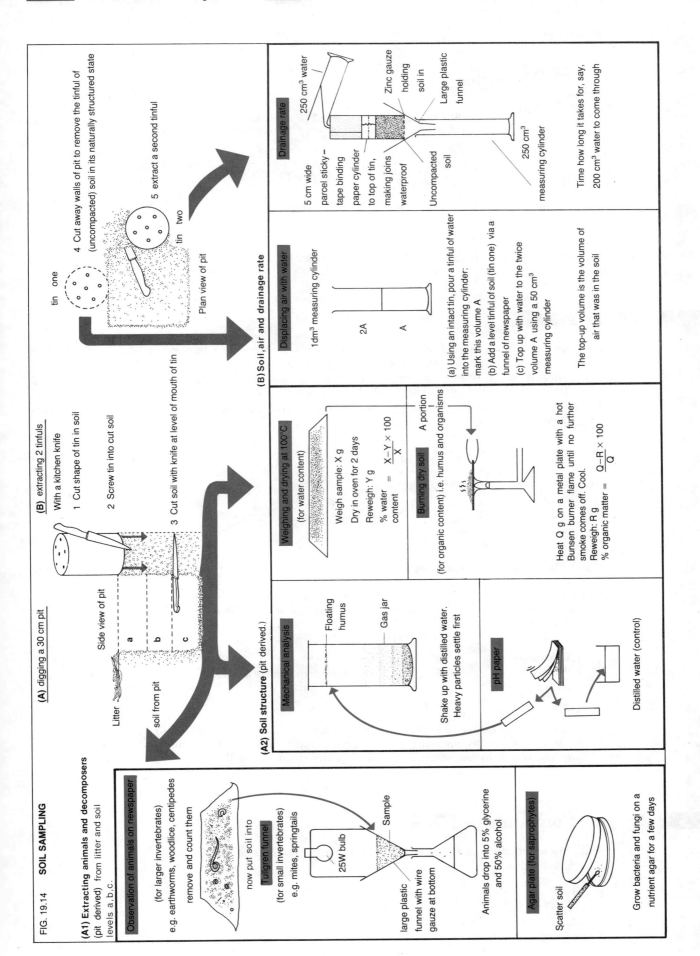

Fig. 19.14 Some methods of studying soil

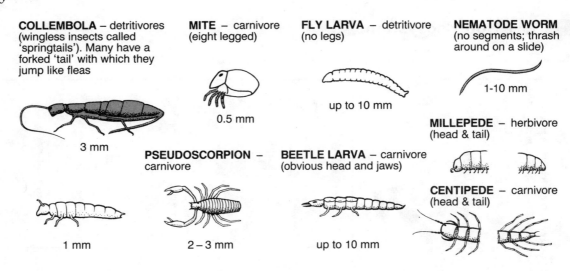

Fig. 19.15 Organisms commonly extracted by Tullgren funnel

19.8 Keys

Keys are a means of identifying organisms in *local* situations, e.g. in pond or woodland. The user of the key selects one of two *contrasting* descriptions, choosing the one that fits the organism being identified. The chosen description leads to a number, alongside which are further descriptions from which to choose. The final choice leads to the organism's name.

Example: Choose one of the organisms in Fig. 19.16 and use the key below the diagram to identify it.

Fig. 19.16 A variety of organisms

| With wings | **1** | (*Now look at descriptions by* **1** *below*) |
| Without wings | **2** | (*Now look at descriptions by* **2** *below*) |

1 ⎰ Two legs — C – Bat
   ⎱ Six legs — B – Butterfly

2 ⎰ With legs — A – Woodlouse
   ⎱ Without legs — **3** – (*Now look at description by* **3** *below*)

3 ⎰ With eyes — D – Fish
   ⎱ Without eyes — E – Earthworm

In the example above, use of internal characteristics (e.g. vertebrae) or confusing ones (e.g. hairiness) would delay identification – some butterflies are as hairy as bats!

Your key of the organisms above could be different but still be 'correct' – if it works.

19.9 Soil Components

Soil: the layer of earth that contains organisms. Consists of
1 rock particles
2 air
3 water
4 mineral salts
5 humus
6 organisms.

Table 19.2 Comparison of sandy soil and clay soil

	Sandy soil	*Clay soil*
1 **Rock particle sizes**	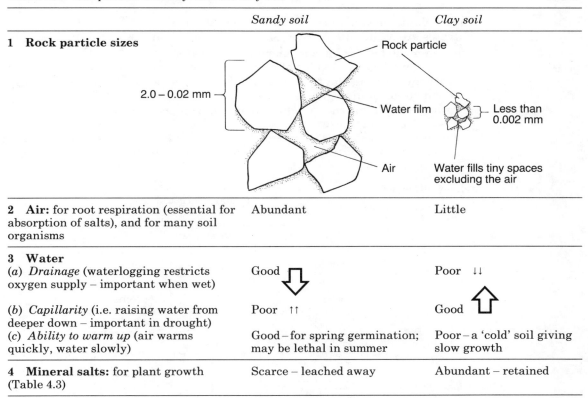	
2 **Air:** for root respiration (essential for absorption of salts), and for many soil organisms	Abundant	Little
3 **Water** (*a*) *Drainage* (waterlogging restricts oxygen supply – important when wet)	Good ⇩	Poor ↓↓
(*b*) *Capillarity* (i.e. raising water from deeper down – important in drought)	Poor ↑↑	Good ⇧
(*c*) *Ability to warm up* (air warms quickly, water slowly)	Good – for spring germination; may be lethal in summer	Poor – a 'cold' soil giving slow growth
4 **Mineral salts:** for plant growth (Table 4.3)	Scarce – leached away	Abundant – retained

Thus a **silty soil,** with its particle size intermediate between sand and clay (0.02–0.002 mm) 'averages out' their properties, giving near-ideal conditions for plant growth.

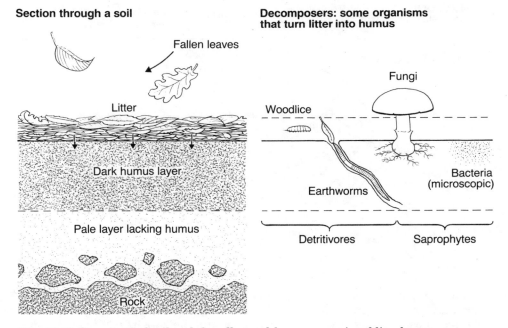

Fig. 19.17 Structure of soil and the effects of decomposers in adding humus

5 Humus – dead organic matter in soil. Mostly decomposing *litter* (fallen leaves) *or manure.* Improves soil by: (*a*) providing mineral salts from decay; (*b*) providing air spaces (improves clay); (*c*) retaining moisture (improves sand); (*d*) improving crumb structure (prevents soil from being blown away); (*e*) encouraging earthworms (see unit 19.12).

6 Organisms: assist circulation of elements. **Decomposers** in soil are particularly important, turning dead organic matter (unusable by green plants) into inorganic food for them, e.g. salts and CO_2. **Bacteria** have special roles in the circulation of nitrogen and carbon in nature (see Fig. 19.18). Roots of **green plants** bind soil, preventing erosion. De-forestation, particularly on slopes, allows rain to wash soil away; over-grazing, especially by goats, allows wind or rain to remove soil.

19.10 Nitrogen Cycle

Green plants need nitrates for protein synthesis.

Nitrates are available to green plants from four sources:

(*a*) man-made fertilizers, e.g. ammonium nitrate;

(*b*) lightning – causes oxides of nitrogen to form; these become nitric acid in the rainfall.

(*c*) **nitrogen-fixing bacteria** – the only organisms capable of converting nitrogen gas into compounds of nitrogen;

(*d*) **nitrifying bacteria** – oxidize ammonium compounds to nitrites and then nitrates, if there is air for them to use.

Nitrates are turned into nitrogen by **de-nitrifying** bacteria if the soil lacks air, as in water-logged conditions. Nitrogen gas is useless to green plants.

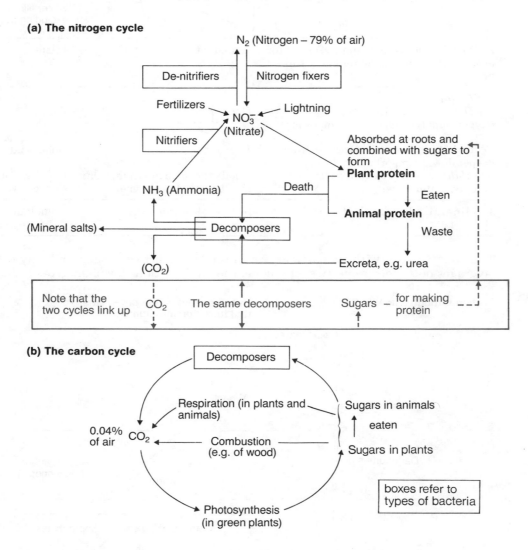

(a) The nitrogen cycle

(b) The carbon cycle

Fig. 19.18 (a) The nitrogen cycle   (b) The carbon cycle

In green plants, nitrates and sugars form amino-acids; these become proteins. Animals convert plant proteins into their own, but in doing so waste some, e.g. as urea, which is excreted.

Decomposers break down dead organisms and their wastes. Nitrogen compounds in them, e.g. proteins, end up as ammonia and then ammonium compounds.

19.11 Carbon Cycle

Green plants *photosynthesize* CO_2 into sugars. Most other organic molecules are made using sugar, e.g. cellulose in wood or proteins and oils in seeds and leaves. When these are eaten by animals, the digested products are turned into animal carbohydrates, fats and proteins.

This variety of organic molecules is returned to air as CO_2 during respiration in plants, animals and bacteria of *decay*; or by *combustion*.

Fuels include wood and the 'fossil fuels' coal, petroleum and natural gas. Fossil fuels were formed by the partial decay and compression of plants by earth-forces millions of years ago.

19.12 Earthworms and Soil

Earthworms (*see also* unit 21.4):
(*a*) *aerate* and *drain* soil by tunnelling;
(*b*) *fertilize* soil by:
(i) pulling litter down into tunnels for bacteria to decompose
(ii) excreting urine
(iii) decomposing when dead;
(*c*) bring *salts*, leached to lower layers, up again to roots (in worm casts);
(*d*) *neutralize* soil acidity by secreting lime into it (from gut glands);
(*e*) *grind* coarse soil finer in gut (in gizzard).

   These activities are exactly what a farmer aims to do to make a *loam* (cultivated soil, with all six soil components in proportions suitable for good plant growth).

19.13 Water Cycle

In the *water cycle* most of the water circulated does not go through organisms (Fig. 19.19).

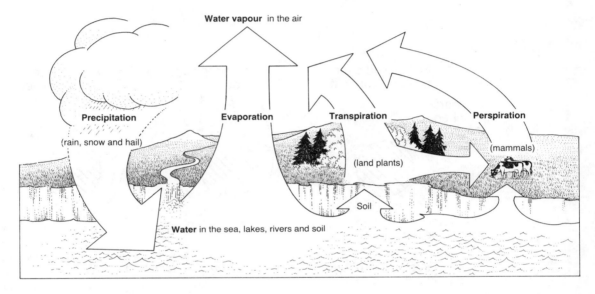

Fig. 19.19 The water cycle

   It has long been suggested that cutting down forests decreases rainfall in that area.

20 MAN AND HIS ENVIRONMENT

Man has two environments: that outside his skin and the other inside it. He must manage both if he is to stay alive as an individual and as a species.

MAN'S EXTERNAL ENVIRONMENT
   Man produces more food for himself than nature alone could provide, by farming the land. Farming depends on producing a fertile soil (see unit 19.9); on breeding good plant and animal food-species (see unit 18.6); and on reducing their pests and diseases (see unit 20.4).

Agricultural practices
1 Ploughing
2 Liming
3 Manuring (= fertilizing)
4 Crop rotation
5 Pest control

20.1 Ploughing

(a) aerates and drains soil by creating ridges and furrows (thus discouraging denitrification);
(b) brings leached salts up to near the surface for roots;
(c) brings pests, sheltering deep down, up to the surface for frost to kill;
(d) allows frost to break up the ridges of soil;
(e) turns organic matter, e.g. wheat stubble, into the ground to decay.

20.2 Liming and Fertilizing

Liming – addition of powdered $CaCO_3$:
(a) neutralizes acidity;
(b) allows efficient application of fertilizers (see Fig. 20.1);
(c) flocculates ('clumps') clay particles together into larger groups ('crumbs') with air spaces between them.

Fertilizing – restoring mineral salts (lost in crops) to the soil.

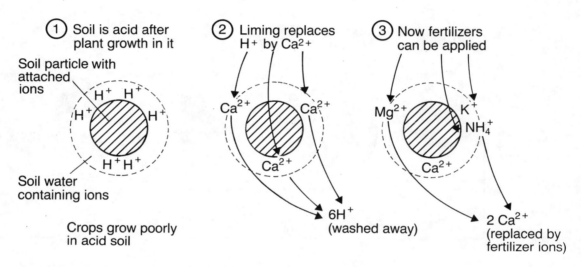

Fig. 20.1 Liming and fertilizing – chemical effects

Table 20.1 Comparison of organic and inorganic fertilizers

	Organic: e.g. clover ploughed in to decay; and animal dung + urine	**Inorganic:** factory products, e.g. $(NH_4)_2SO_4$ or wastes, e.g. basic slag
Cost	Cheap	Expensive
Application	Difficult (bulky, sticky)	Easy (powders, granules)
Action	Slow but long-lasting	Quick but short-lasting
Soil structure	Improved (see 'humus')	Not improved
Earthworms	Encouraged	Can harm them

20.3 Crop Rotation

Crop rotation – growth of different crops on the same land in successive years without manuring each year. The two harvested crops have different mineral requirements and often obtain them from different soil depths. In the 'fallow year' legumes, e.g. clover, are sown to restore *nitrogen compounds* to the soil when the plants decay after being ploughed in. Other minerals (removed in crops) are restored by fertilizing. (See Fig. 20.2.)

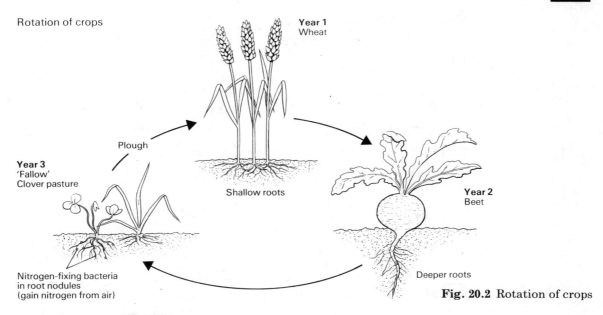

Rotation of crops

Year 1
Wheat

Plough

Year 3
'Fallow'
Clover pasture

Shallow roots

Year 2
Beet

Deeper roots

Nitrogen-fixing bacteria
in root nodules
(gain nitrogen from air)

Fig. 20.2 Rotation of crops

20.4 Pest Control

Pest organisms, e.g. locusts, termites, weeds, reduce man's agricultural efforts or other interests. It was *Man* himself who created pests by providing organisms, normally held in check in their ecosystems, with unusual opportunities for increase in numbers in a monoculture (e.g. of corn, cotton or cows).

Chemical control – expensive. May eliminate pest but also kills harmless organisms too. Examples: DDT (insects); 2–4D (weeds) (see Table 20.3).

Biological control – cheap. Use of a natural enemy of the pest to *control* numbers (some damage must be expected), e.g. guppy fish eat mosquito larvae in ponds (see also unit 21.12).

20.5 Human Population Crisis (Problems)

Man's population growth has been *exponential:*

Year	Population	
1630 (estimated)	500 million	200 years
1830	1000 million	100 years
1930	2000 million	
1975	4000 million	estimated to rise to 6000 million by 2000 (see Fig. 16.10)

This has occurred because of improvements in:

1 **Agriculture:** more food per hectare owing to improved strains of crops and livestock (see unit 18.6); mechanization and fertilizers.

2 **Sanitation:** disposal of excreta, finally via sewage farms.

3 **Water supply:** filtration, finally chlorination.

4 **Medicine:** inoculation, drugs, antibiotics, aseptic surgery.

reduced death rate from disease

Consequences

Bacteria on an agar plate show exponential growth in population. This leads to exhaustion of food and self-pollution resulting in mass death (crash phase, see Fig. 16.8). Similarly, Man's exponential growth in population is resulting in both pollution of the biosphere and reduction of its resources. Unlike bacteria, Man has the ability to avoid the crash phase by using a variety of solutions.

20.6 Pollution

Pollution: waste substances or energy from human activities which upset the normal balances in the biosphere. Anything from noise (aircraft) and heat (atomic power stations) to various substances in excess (sewage, DDT).

Table 20.2 Air pollutants

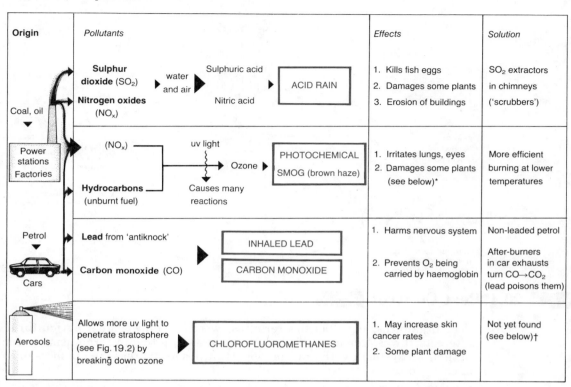

Laws: the Clean Air Acts (1956 and 1968) have prevented the sooty smogs of industrial and city areas. Smogs caused death by bronchitis and traffic accidents, and dirtied and damaged buildings of stone. Britain (1985) reduced lead in petrol from 0.4 to 0.15 g/dm³. In some other EEC countries lead-free petrol is in widespread use.

* *Forest damage* in Europe and the US has been blamed on 'acid rain'. This pollution raises aluminium levels (toxic) and removes calcium, magnesium and potassium (nutrients) from soil; but it also kills lichens. Yet trees are severely damaged in some areas where lichens flourish and where SO₂ levels are well below those known to cause damage. In contrast to steadily falling SO₂ levels in Europe, ozone levels are rising and can reach high concentrations in rural as well as industrial areas. Ozone damages chlorophyll. Unknown diseases may be the cause since individual trees flourish in otherwise devastated areas (possibly mutants? – see unit 17.15). No single cause for tree damage has been established.

† Industrial use of chlorofluoromethanes is increasing. In Antarctica a progressive thinning of the ozone layer (by over 50% in October 1985) has been detected each winter since 1979. Some thinning is said to occur over N. Europe.

Table 20.3 Land and water pollutants

Land pollutants	*Origin*	*Effect*	*Solution*
(a) Insecticides	Crop protection; control of disease vectors, e.g. mosquito	May kill top consumers*; may lower photosynthesis rate of marine algae	Ban undesirable ones, e.g. DDT, as UK has done†
(b) Radioactive wastes	Nuclear reactor accidents and wastes; atom bombs	Mutations	Nuclear waste silos – but some have leaked
Water pollutants			
(a) Sewage	Human	Eutrophication §	Sewage treatment (see unit 3.12)
(b) Artificial fertilizers	Excessive agricultural use	Eutrophication §	Use of organic manures (unit 20.2)
(c) Petroleum	Tanker accidents	Oiled sea birds, beaches⊕	Effective accident prevention
(d) Mercury (organic)	Chemical works; fungicides on seeds, wood	Minamata disease (paralysis, idiots born)*	Effluent purification

*Tiny amounts in producers are concentrated along a food chain into top consumers. Thus in the 1950s eagles had very high DDT levels and laid thin-shelled eggs that broke easily. Their population fell. Similarly, plankton in Minamata Bay, Japan, absorbed small amounts of mercury waste in a factory's effluent. This was concentrated in predators – crustaceans, then fish – and finally got into the human fish-eating population. Many people died or became paralysed. Mothers gave birth to idiot and malformed children.

† Poor countries cannot afford to do this in the tropics – famine or disease would result. DDT is cheap, effective.

§ **Eutrophication:** enrichment of natural waters with mineral salts.
(a) *Mineral salts* drain off recently over-fertilized fields; or are formed by bacterial breakdown of sewage in the water.
(b) *Algae multiply* exponentially, given this excess of food (water goes green). They crash in numbers as their food runs out.
(c) *Bacteria multiply* exponentially, given an excess of dead algae to decay.
(d) Bacterial respiration sharply *reduces* O_2 in the water – *killing aquatic animals* by suffocation. This worsens the situation as even more bodies decay.

⊕ Detergents have a harmful effect on cell membranes and affect aquatic life seriously. Detergents often harm marine life more than the oil slicks they are used on (to disperse them). Older-type detergents used to release phosphates on break-down – encouraging eutrophication. Modern detergents do not and they break down easily.

20.7 Depletion of Resources

Resources are of two types: non-renewable (non-living) and renewable (living).
(a) *Non-renewable*, e.g. **minerals;** and natural gas which will be used up within 50 years (from *known* sources).
Soil is being lost by erosion because of unwise land use, e.g. over-grazing or clear felling of forest (thus removing the binding action of roots), particularly on sloping land, e.g. in Amazonia, Philippines and the Himalayas. New soil takes centuries to form through weathering of rock and the action of organisms (see unit 19.9). Without fine soil, many producers cannot grow and whole ecosystems may be destroyed.
(b) *Renewable*, e.g. **foods:** herring and whales have been over-fished. Cutting down of forests exceeds planting. Harvesting should not exceed replacement rate.

Destruction of wild-life
Agricultural needs destroy natural habitats; pesticides, poison; and hunting for 'sport' or fashionable items, e.g. skins, ivory, may all lead to extinction of species, e.g. dodo, Cape lion.

Food shortage
Two thirds of the world population lack either enough energy-foods or protein or both in their diet. Poor nations are unable to pay for the surplus food of rich ones.

Reduced living space
Overcrowded populations lead to greater chance of epidemic diseases and social diseases, e.g. vandalism, child abuse, drug-taking, alcoholism. In Britain every year about 10 000 children are registered as abused, of whom about 200 die.

20.8 Human Population Crisis (Solutions)

Solutions

1 Contraception (see unit 15.4) and abortion (removing unwanted embryos) would by themselves reduce the rate of increase in population if used world-wide. Some people object to these practices.

2 Conservation of minerals: use of substitutes for metals, e.g. carbon-fibre plastics; reversing the throw-away mentality by making durable products, e.g. cars that last; recycling metals in discarded items. (see also unit 3.12).
Note: Lowered industrial production (and fewer jobs) must be accepted as a result of these policies.

3 Conservation of wildlife and natural scenery: strict guardianship of nature reserves; acceptance that minerals in a mountain may be less ,valuable than the beauty it affords. Man's need for recreation and enjoyment of nature is as necessary for health as meeting his material needs. Reclamation of gravel pits and of mining tips by suitable planting and landscaping can provide amenity areas, e.g. for sailing and as parks.

4 Conservation of genes: wild animals and plants may not be of direct use to Man but can provide useful genes for introduction into his breeding programmes, e.g. genes for disease resistance from the small inedible potato of South America, genes for hardiness in Soay sheep, and genes for high vitamin C content in wild tomatoes.

5 Conservation of renewable resources: by never taking more than can be replaced (by reproduction). Re-forestation is thus a priority world-wide. Meanwhile recycling paper helps reduce tree-felling.

6 New sources of food: greater dependence on micro-organisms, e.g. 'SCP' and mycoprotein (see unit 3.12), and soya bean meat-substitutes. Farming wild animals on their natural land supplemented by cattle feedstuffs produces high quality lean meat, fast, e.g. red deer (Scotland), wildebeeste (Africa).

7 Finding new (acceptable) energy sources, e.g. solar power, tide power. Fast-breeder nuclear reactors will produce very much more dangerous waste than conventional reactors – a possible mutation hazard. But using fossil fuels (coal, oil) to a greater extent may raise CO_2 levels in air causing a *greenhouse effect* that raises global temperature. This could melt polar ice caps, thus raising sea level and flooding many major coastal cities, e.g. London, New York. To recycle metals, produce substitutes for them, and make artificial fertilizers is very energy-consuming.

MAN'S INTERNAL ENVIRONMENT
Hormones and nerves (see unit 12.11) help to stabilize the body's internal environment (achieve homeostasis). Any change from normal is called **disease.**

20.9 Types of Disease in Man

1 Genetic: since these diseases are inherited, they are *incurable.*
Examples: **haemophilia** – a gene mutation; **mongolism:** baby has an extra chromosome, i.e. 46 + 1 owing to abnormal meiosis in the mother. Person has retarded development and dies usually before the age of 40 (see unit 17.15).

2 Diet deficiency: curable by eating a balanced diet.
Examples: lack of iodine (**goitre,** see unit 12.10); or vitamin C (**scurvy,** see unit 4.5); or protein (**kwashiorkor** – matchstick limbs, pot belly, see unit 4.6).

3 Hormonal: curable by artificial supply of hormone.
Examples: lack of thyroxine (**cretinism**) or insulin (**diabetes**) (see unit 12.10).

4 Pathogenic: entry of parasites (pathogens) into body which upset its metabolism.
Examples: viruses (see unit 3.1); bacteria (see unit 3.2); protozoa (see unit 3.11).

(*a*) For **prevention** (better than cure):
(i) **kill vectors,** e.g. mosquitoes carrying malaria; or intermediate hosts, e.g. snails carrying bilharzia;
(ii) **prevent access** of parasite by hygiene, water chlorination, cooking food or protective measures, e.g. mosquito nets;
(iii) employ **preventive medicine** (prophylaxis) using immunization (see below) or drugs, e.g. mepacrine for malaria;
(iv) **quarantine** those who are ill (isolate sources of infection).
(*b*) A **cure** requires:
(i) **hospitalization:** rest and good food assist body's own defences;
(ii) **medicines:** drugs, antibiotics kill pathogens.

5 Environmental: often preventable by wise precautions.
Examples: skin cancer (excessive sunbathing); industrial chemicals (not wearing face masks).

20.10 Natural Defences of the Body Against Pathogens

1 Skin: keratin; sweat (which is antiseptic) – (see unit 10.10).
2 Blood clotting: provides a temporary barrier before wound heals (see unit 8.2).
3 Phagocytes: ingest microorganisms (see unit 8.2).
4 Lymphocytes: make antibodies (see unit 8.2) to kill pathogens or neutralize their poisons (with anti-toxins), thus making the body immune (protected). There are two methods of immunization:

Table 20.4 Comparison of active and passive immunity

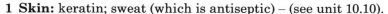

	Active immunity (body participates)	**Passive immunity** (body passive)
Method	Weakened or dead strain of pathogen introduced, e.g. polio **vaccine**	Antibodies made by another animal, e.g. horse, or genetically engineered, are injected (see unit 3.12)
Protection	(*a*) long-lasting ('boosters' prolong protection, e.g. anti-tetanus every 3 years) (*b*) takes weeks to develop	(*a*) short-lived (body destroys the foreign antibodies) (*b*) immediate protection

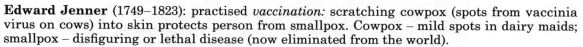

20.11 Notable Contributors to Health and Hygiene

Edward Jenner (1749–1823): practised *vaccination:* scratching cowpox (spots from vaccinia virus on cows) into skin protects person from smallpox. Cowpox – mild spots in dairy maids; smallpox – disfiguring or lethal disease (now eliminated from the world).

Louis Pasteur (1822–1895): father of *bacteriology*. Discovered the bacterial nature of putrefaction and many diseases. Saved silk industry (pebrine disease of silkworms), brewers ('ropy' beer), poultry farmers (chicken cholera) and cattle farmers (anthrax) from severe losses by developing sterile techniques and vaccines. Finally, developed a rabies vaccine to protect humans.

Joseph Lister (1827–1912): developed *antiseptic surgery*. Used fine phenol spray to kill bacteria during operations, dramatically reducing hospital deaths. Today *aseptic* surgery is used – sterilization of all equipment before use, in autoclaves (see unit 3.5).

Alexander Fleming (1881–1955): discovered lysozyme (natural antiseptic in tears and saliva) and the *antibiotic* penicillin (see unit 3.9).

DRUGS, ANTIBIOTICS, DISINFECTANTS AND ANTISEPTICS

'Drugs' are chemicals made by Man or organisms. Some are harmful and possession of them is illegal, e.g. LSD – which has no medical purpose. Others (in the right doses) assist medically, e.g. sulphonamides for curing bacterial infections, aspirin for headaches, and belladonna for helping people with ailing hearts. The term 'drug' is thus too vague to be very useful.

Antibiotics are chemicals secreted by bacteria or fungi and extracted by Man for his own use in killing microorganisms in his body. *Examples:* penicillin, aureomycin. Accurate choice of antibiotic to treat an infection depends on taking a swab (see unit 3.6).

Disinfectants are chemicals made by chemists to kill microorganisms, e.g. neat 'Dettol' in toilets.

Antiseptics are chemicals used in such a dose that they kill microorganisms but *not* human cells with which they make contact. May be diluted disinfectants, e.g. weak 'Dettol' for gargling or bathing cuts.

20.12 Options for a Human Future

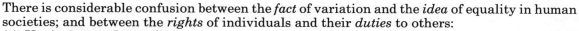

There is considerable confusion between the *fact* of variation and the *idea* of equality in human societies; and between the *rights* of individuals and their *duties* to others:
(*a*) **Variation and equality**
Human races are different; so are individuals and males and females. Biologically speaking each individual has different adaptations, i.e. potential. Yet in the Declaration of Independence of the United States (1776) is the statement 'We hold these truths to be self-evident, that all men are created equal'.
(*b*) **Rights and duties**
Humans like to have the freedom to 'do their own thing' – to show their natural variation through behaviour. When humans live close together this is not entirely possible – whether in a family, a town or in a family of nations. A large human population in the world results in actions by some people that strongly affect others. Such actions would probably have gone unnoticed in the past when the population was small since the effect was small, e.g. the discharge of sewage into rivers.

As the population rises, the **rights** of individuals become reduced because their **duties** to others (to preserve *their* rights) increase. The 'right' to manufacture goods or to provide energy must be balanced by the 'duty' to minimize pollution (see unit 20.6). If necessary, governments are forced to impose laws restricting the 'rights' of some to preserve the 'rights' of others. The 'right' to have as many children as a couple wishes may have to be taken away in some countries – as it has been in China.

Here are some options for humans to think about:

1 **Voluntary population control** (see unit 15.4)
With sensible population size in each country, demands for food, materials and energy would be reasonable; and pollution could be controlled. Humans could live in balance with the biosphere.

2 **Continued population increase**
This might result in
(*a*) famine: through misuse of land for agriculture;
(*b*) disease: through greater likelihood of epidemics; breakdown of health services;
(*c*) war and civil unrest: see no. 3.
The fate of deer on the Kaibab plateau (see unit 18.2E) is warning enough of what might happen to a population too large for available resources.

3 Preparation for success in competition

Countries or races may decide to compete for the available resources – a return to the laws of natural selection and a breakdown of human society. In nature a high rate of reproduction is an adaptation for survival and success, e.g. rabbits, locusts. With human societies it is a recipe for disaster – the population remains in poverty and is technologically underdeveloped. The technological means of eliminating competitors (weaponry) has gone far beyond the club and spear: this may affect survivors as well.

Malthus (see unit 18.4) predicted that no. 2 above would control human population. He also hoped for solution no. 1, but he had no knowledge of modern means of contraception (see unit 15.4).

Which is to be the human solution?

21 A VARIETY OF LIFE

21.1 Algae

(See also units 2.3 and 3.10.)

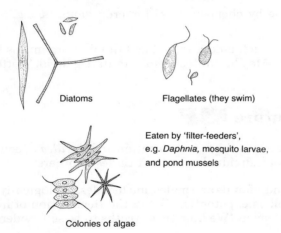

Diatoms

Flagellates (they swim)

Eaten by 'filter-feeders', e.g. *Daphnia,* mosquito larvae, and pond mussels

Colonies of algae

Fig. 21.1 A selection of algae commonly found in fresh water as plankton

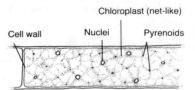

Fig. 21.2 Part of a cell of *Cladophora,* a *branching* filamentous alga forming dense green 'blankets' in ponds – compare *Spirogyra* (unit 3.10)

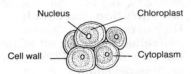

Fig. 21.3 *Pleurococcus,* 'green dust' on the wettest side of tree-trunks, posts, etc

21.2 Mosses and Ferns

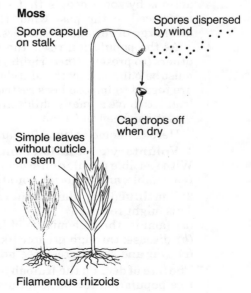

Moss

Spore capsule on stalk

Spores dispersed by wind

Cap drops off when dry

Simple leaves without cuticle, on stem

Filamentous rhizoids

Fig. 21.4 Moss plant
Most mosses do not survive dry conditions: Leaves one cell thick without waterproofing cuticle; shallow rhizoids not able to gather much water.
However wall-mosses can survive drying – recover with rain

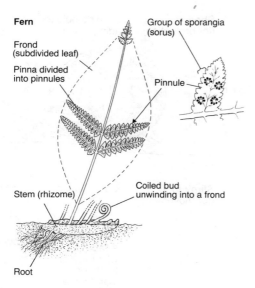

Fig. 21.5 Fern plant
Ferns can survive dry conditions.
Leaf structure similar to that of flowering plants; roots gather water well; good transport system for water and food; stem stores food

Both mosses and ferns require wet conditions for sexual reproduction. Thus neither can colonize very dry habitats.

21.3 Flowering Plants

See also units 14.3 (winter twig), 5.5 (leaf), 14.5 (flowers), 14.10 (fruits).

Fig. 21.6 General structure of a dicot plant

21.4 Annelids

See also unit 2.4.

The earthworm is a scavenger (detritus feeder) in soil – see unit 19.7.

Eaten by many birds, e.g. thrush; hedgehogs, foxes.

In ponds, *Tubifex* worms (about 1 mm in diameter) live in masses in tubes that they make in the mud. When water has less O_2, e.g. in warm weather, the worms partly emerge, actively waving around in the water like moving hair to get O_2. Also a detritus feeder.

Eaten by fish.

Both earthworms and *Tubifex* have haemoglobin in their blood to assist oxygen uptake in their relatively airless habitats.

21.7 Earthworm (*Lumbricus*)

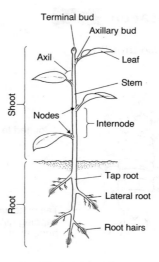

21.5 Molluscs

See also unit 2.4.

The land snail is a herbivore, feeding on ground-layer plants. Eaten by thrushes, hedgehogs.

Pond snails, e.g. *Limnaea* and *Planorbis*, have only one pair of tentacles, with eyes at their base.

Conserve moisture by:
1 *Activity* only in damp or shady conditions (or at night).
2 *Mucus* (slime) to slow down evaporation.
3 *Lung* to reduce water loss (compared with gills).
4 *Aestivating* in dry conditions: animal withdraws into its waterproof shell and seals the opening with mucus. This hardens into a waterproof membrane (the epiphragm).

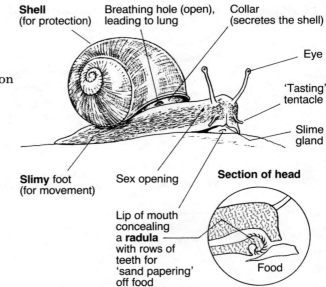

Fig. 21.8 *Land snail*, e.g. *Helix* (garden snail) or *Cepea* (woodland and grassland snail)

21.6 Crustacea

See also unit 2.4.

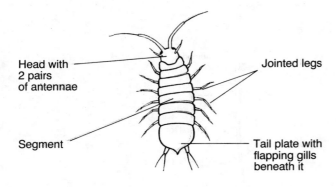

Fig. 21.9 *Asellus* (water-louse)
A scavenger (detritus-feeder) in fresh water, like the woodlouse on land. Eaten by fish.

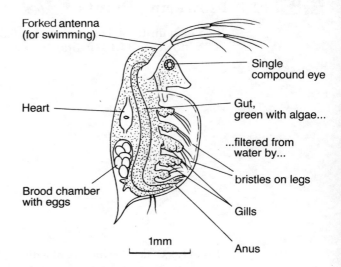

Fig. 21.10 *Daphnia* (water-flea)
A herbivore; part of the plankton in fresh water. Feeds on diatoms and other algae. Eaten by fish.

21.7 Insects

Insect characteristics

As insects are **Arthropods** (unit 2.4) they also have:
1 an exoskeleton of chitin;
2 discontinuous growth – see below (compare humans, unit 16.3);
3 many-jointed legs;
4 segmented body.

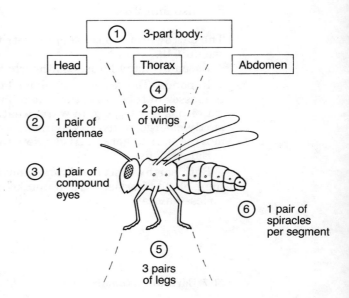

Fig.21.11 Six insect characteristics.

Growth is discontinuous. The exoskeleton does not grow. It has to be shed (**ecdysis**) from time to time.

To do this the old skeleton is partly digested away and finally split open. The insect emerges with a new soft exoskeleton, which it expands before it hardens at a larger size, within an hour. So the size of an insect increases in bursts (see Fig. 21.12). Each stage between ecdyses is called an **instar**.

Fig. 21.12 Discontinuous growth pattern of an insect

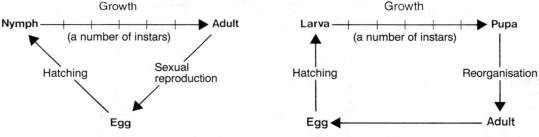
(Arrows indicate ecdyses)

Two kinds of life cycle:

1 Incomplete metamorphosis	2 Complete metamorphosis
Examples: locust, cockroach, dragonfly	butterfly, bee, (most insects)
Growing stage: **nymph** (similar to adult, lacking only wings and ability to reproduce). Last ecdysis gives adult.	**larva** (so unlike adult that reorganization into an adult must be achieved as a **pupa**).

Fig. 21.13 Two kinds of life cycle in insects

Incomplete metamorphosis cycle:
Nymph → (Growth, a number of instars) → Adult → (Sexual reproduction) → Egg → (Hatching) → Nymph

Complete metamorphosis cycle:
Larva → (Growth, a number of instars) → Pupa → (Reorganisation) → Adult → Egg → (Hatching) → Larva

Metamorphosis is the change from a young to an adult form. Young stages in insects have no *wings*; nor can they *reproduce*.

21.8 Locust

The desert locust (*Schistocerca gregaria*) – found from North Africa to India. Devastates vegetation of all kinds, both as 'hopper' (nymph) and adult. Controlled by laying bran, soaked in insecticide, in path of hoppers (see Fig. 21.14). If necessary, flying swarms of adults must be sprayed with insecticide from planes.

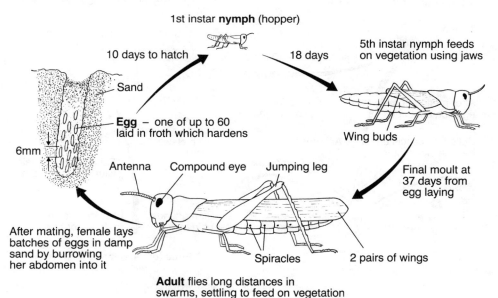

1st instar **nymph** (hopper)

10 days to hatch   18 days

5th instar nymph feeds on vegetation using jaws

Wing buds

Final moult at 37 days from egg laying

Egg – one of up to 60 laid in froth which hardens

Sand

6mm

Antenna   Compound eye   Jumping leg

After mating, female lays batches of eggs in damp sand by burrowing her abdomen into it

Spiracles   2 pairs of wings

Adult flies long distances in swarms, settling to feed on vegetation

Fig. 21.14 The locust, *Schistocerca gregaria* (showing incomplete metamorphosis)

21.9 House-fly

House-fly (*Musca domestica*).

Adults transmit diseases (e.g. dysentery, certain worms) by visiting faeces and then human food. Here they deposit the infecting organisms via their feet or saliva (see Fig.6.1) or by their own droppings ('fly-spots'). Flies controlled by good garbage disposal and sanitation (removes breeding sites); insecticides. In tropics: use muslin or wire gauze fly-screens to cover food and drink.

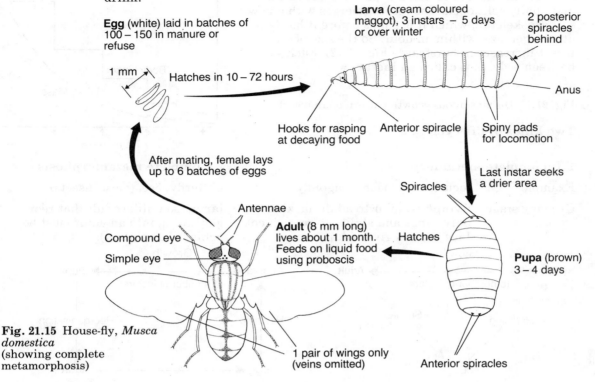

Egg (white) laid in batches of 100 – 150 in manure or refuse

1 mm

Hatches in 10 – 72 hours

Larva (cream coloured maggot), 3 instars – 5 days or over winter

2 posterior spiracles behind

Anus

Hooks for rasping at decaying food

Anterior spiracle

Spiny pads for locomotion

Last instar seeks a drier area

Spiracles

After mating, female lays up to 6 batches of eggs

Antennae

Compound eye

Simple eye

Adult (8 mm long) lives about 1 month. Feeds on liquid food using proboscis

Hatches

Pupa (brown) 3 – 4 days

1 pair of wings only (veins omitted)

Anterior spiracles

Fig. 21.15 House-fly, *Musca domestica* (showing complete metamorphosis)

21.10 Large Cabbage White Butterfly

Large cabbage white butterfly (*Pieris brassicae*) eats cabbage-family plants. Controlled by insecticides and a parasitic 3 mm black wasp (*Apanteles glomerata*). The wasp's eggs, injected into caterpillar, hatch into larvae feeding on caterpillar's tissues, thus killing it. Pupates within bright yellow cocoons on caterpillar's skin.

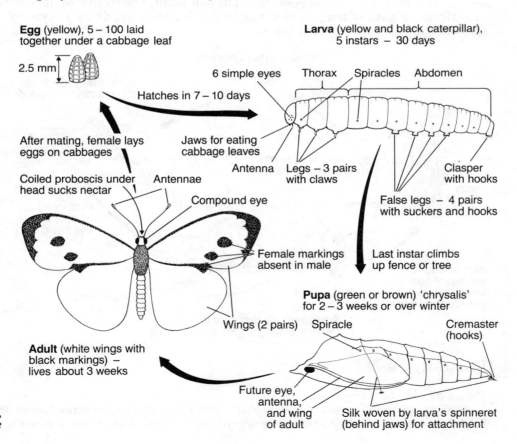

Egg (yellow), 5 – 100 laid together under a cabbage leaf

2.5 mm

Larva (yellow and black caterpillar), 5 instars – 30 days

Hatches in 7 – 10 days

6 simple eyes   Thorax   Spiracles   Abdomen

After mating, female lays eggs on cabbages

Jaws for eating cabbage leaves

Antenna   Legs – 3 pairs with claws

Clasper with hooks

Coiled proboscis under head sucks nectar

Antennae

Compound eye

False legs – 4 pairs with suckers and hooks

Last instar climbs up fence or tree

Female markings absent in male

Pupa (green or brown) 'chrysalis' for 2 – 3 weeks or over winter

Wings (2 pairs)   Spiracle

Cremaster (hooks)

Adult (white wings with black markings) – lives about 3 weeks

Future eye, antenna, and wing of adult

Silk woven by larva's spinneret (behind jaws) for attachment

Fig. 21.16 Large cabbage white butterfly, *Pieris brassicae*

21.11 Honey bee

Honey bee (*Apis mellifera*) (see Fig. 21.17).

Organization in the hive
No individual bee can live for long without assistance from the others. Thus the hive, with its 5000–100 000 bees, is comparable to a socially-organized unit, e.g. a town.

The queen is the only fertile female (*a*) laying eggs and (*b*) secreting 'queen substance', which is passed from bee to bee by mouth and keeps the colony working together.

Drones are fertile males; do no hive work; they are fed by workers and driven out to die in autumn.

Workers are infertile females with a sequence of duties as they get older. Their duties include:
(*a*) nursing – look after larvae, feeding them 'royal jelly' (from head-glands), honey and pollen;
(*b*) hive maintenance – make honey-comb (from wax secreted by abdomen) and store pollen and honey in it;
(*c*) foraging – gather nectar from flowers and turn it into honey in crop;
(*d*) defending (by sting) and ventilating (by wing-beating) the hive.

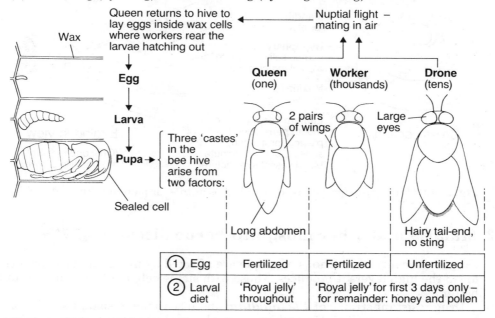

		Queen (one)	Worker (thousands)	Drone (tens)
①	Egg	Fertilized	Fertilized	Unfertilized
②	Larval diet	'Royal jelly' throughout	'Royal jelly' for first 3 days only – for remainder: honey and pollen	

Fig. 21.17 Life cycle of the honey bee, a social insect

21.12 Mosquito

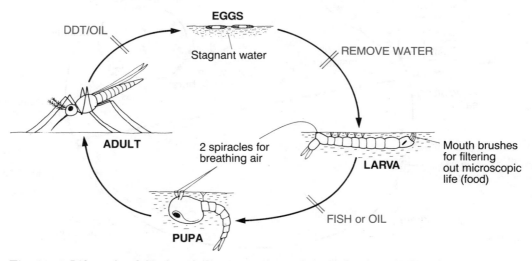

Fig. 21.18 Life cycle of the Anopheline mosquito and methods of controlling it

The mosquito lays its eggs on stagnant water. The aquatic larvae breathe air through spiracles at the surface of the water; so do the pupae. The adults may emerge within a week after egg-laying in tropical countries. The females need a meal of blood to ensure proper egg development before fertilization. They tend to 'bite' humans at night, sheltering by day in dark places in houses. These habits give opportunities for **controlling mosquitoes:**

1 drain marshes or otherwise remove stagnant water (prevents egg-laying);
2 spray light oils containing insecticide on water that cannot be removed (oils block spiracles, suffocating the aquatic stages; the insecticide kills females landing to lay eggs);
3 introduce 'mosquito fish', e.g. *Gambusia* or guppy, into the water (to eat larvae and pupae);
4 spray walls of houses with long-lasting insecticides, e.g. DDT (kills adults sheltering there).

Mosquitoes only transmit **diseases**, e.g. malaria, if they are given the opportunity to suck up the parasites of an infected person. When mosquitoes 'bite' they inject saliva to prevent the blood clotting. It is with this saliva that the parasites enter healthy people. It is therefore wise in the tropics to:

1 sleep under a mosquito net (prevents getting 'bitten');
2 take, regularly, drugs that kill the parasites that do get injected (prophylaxis, see unit 20.9);
3 quarantine (isolate) those who become diseased, away from other people and under mosquito nets (to prevent infecting mosquitoes).

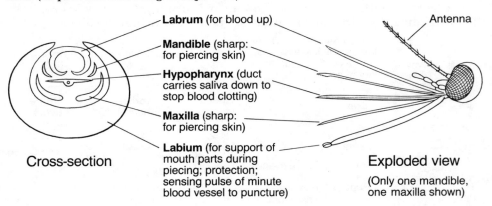

Fig. 21.19 Mouthparts (proboscis) of female mosquito – for piercing skin and sucking blood

21.13 Malaria and other mosquito-borne diseases

Despite precautions, millions of people are affected by **malaria** in the tropics. Large numbers die of the high fevers it produces. Those with sickle-cell anaemia trait (see unit 18.2B) survive better.

Drug resistant strains of ***Plasmodium*** and insecticide-resistant strains of mosquitoes arise by mutation (see unit 17.15), making control of malaria a continuing problem.

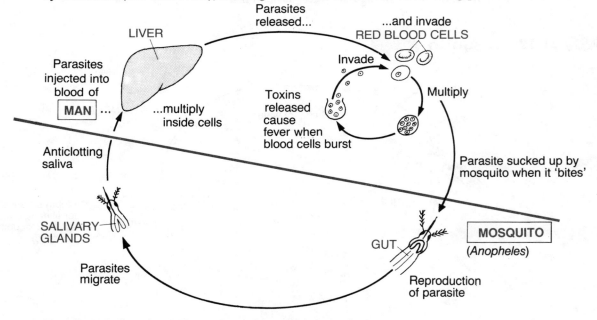

Fig. 21.20 Life cycle of *Plasmodium*, the malaria parasite

Various mosquito species also infect Man with:
Elephantiasis: enormous enlargement of limbs caused by blockage of the lymph vessels by millions of tiny worms (nematodes). Incurable.

Yellow fever: severe jaundice (yellow skin) caused by liver damage from a virus which may cause death.

21.14 Importance of Insects to Man

Helpful

1 Bees: pollinators (without which orchard fruit yields are greatly reduced) and **suppliers of honey** (sweetener) and **beeswax** (for high-grade polishes and lipstick).
2 Biological control of pests, by *ladybirds* (eat aphids, mealy-bugs and scale insects in garden, coffee and citrus plantations); *Cactus moth* caterpillars (eat prickly-pear cactus invading agricultural land).

Harmful

1 Food destroyers, e.g. *locust* (crops), *grain weevil* (stored grain).
2 Material destroyers, e.g. *termites* (wooden buildings), *cotton boll weevil* (cotton flower), *clothes moth* (woollen clothes).
3 Disease vectors, e.g. *mosquitoes* (yellow fever virus, malaria protozoan and elephantiasis nematode worm), *tsetse flies* (human sleeping sickness and similar sicknesses in domesticated animals), *house-fly* (dysentery protozoa and bacteria), *fleas* (plague bacteria), *wood-boring beetle* (Dutch elm disease fungus), *aphids* (plant virus disease).
4 Nuisances, e.g. *cockroaches* and *ants* (spoiling food).

21.15 Pork Tapeworm

An endoparasite of Man and pig. Does little harm to either host unless they are undernourished. Humans have increased appetite; may suffer slightly from tapeworm's toxins.

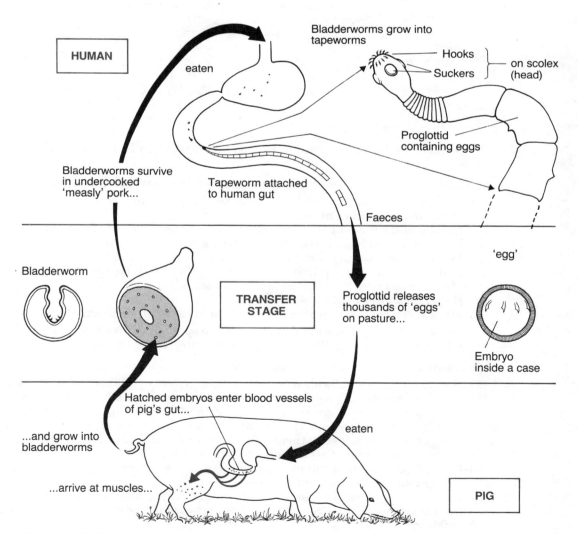

Fig. 21.21 Life cycle of the pork tapeworm, *Taenia solium*

Control – by breaking the life-cycle:
(*a*) *Cook pork thoroughly* – bladderworms are killed by cooking.
(*b*) *Dispose of faeces* sanitarily – pigs cannot be infected.
(*c*) *Inspect pork* – meat inspectors prevent sale of 'measly pork'.

Adaptations to parasitic life
1 **Scolex:** hooks and suckers prevent dislodgement by food flowing in intestine.
2 **Thick cuticle** (perhaps also anti-enzymes): prevent digestion of worm by host.
3 **Flat shape:** large surface area for absorption of food (digested by Man).
4 **Anaerobic respiration:** little O_2 in intestine.
5 **Hermaphrodite and self-fertilizing:** essential because worm is long (2–8 metres), so only one can be accommodated at a time.

21.16 Bony fish

See also unit 2.4.
The three-spined stickleback (*Gasterosteus aculeatus*).

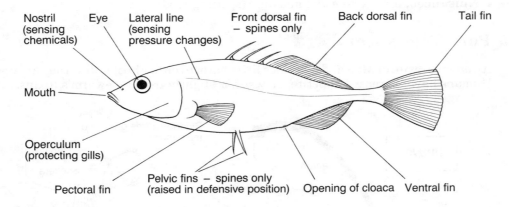

Nostril (sensing chemicals) Eye Lateral line (sensing pressure changes) Front dorsal fin – spines only Back dorsal fin Tail fin

Mouth

Operculum (protecting gills)

Pectoral fin Pelvic fins – spines only (raised in defensive position) Opening of cloaca Ventral fin

Fig. 21.22 Three-spined stickleback

A carnivore feeding on worms, crustacea and insect larvae at pond margins. Eaten by perch, pike and heron.

Reproduction: has complex instinctive courtship behaviour and parental care:
(*a*) In February – March, *males* become red-breasted and blue-eyed and take up a *territory* (defended from other males).
(*b*) Build a *nest*-tunnel of water-weeds stuck together by a kidney secretion.
(*c*) Lead fat, egg-laden females by a *zig-zag dance* to the tunnel.
(*d*) Female enters tunnel and, prodded by male, lays a few *eggs*; she then leaves.
(*e*) Male enters tunnel, squirting eggs with sperm to *fertilize* them.
(*f*) Male *aerates* nest by fin movements and *defends* it.
(*g*) Eggs *hatch* after about a week.
(*h*) Male keeps *fry* together in a defended shoal for another week.

   (Most bony fish lay large numbers of eggs and sperm into the same place in the water, trusting to luck that sufficient fertilization of eggs and survival of the young will take place.)

Adaptation to an aquatic environment
(*a*) *Shape:* streamlined. Skin, secreting mucus, covers overlapping bony scales.
(*b*) *Propulsion:* sideways movement of muscular body exerts a backwards and sideways force on the water via the large surface area of the *tail* (see Fig. 21.23). Muscles on either side contract alternately to give sideways movement. In fast swimming, side fins are kept flat against body. When still, thrusts from pectoral (and pelvic) fins adjust position.
(*c*) *Stability: fins* prevent roll, pitch and yaw (see Fig. 21.24).
(*d*) *Control:* pectoral and pelvic fins act as *hydroplanes* according to angle; when both are held at right angles to body, act as *brakes*.
(*e*) *Buoyancy: air-bladder* (contents adjustable) keeps fish at required depth. Saves energy (cartilaginous fish, e.g. sharks, have no air-bladders and must keep swimming to prevent sinking).
(*f*) *Gills:* for gaseous exchange (see unit 9.7).

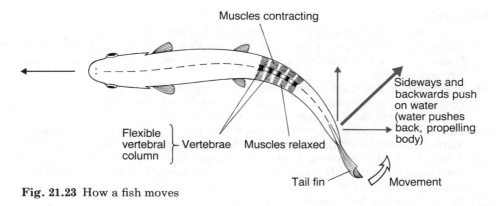

Fig. 21.23 How a fish moves

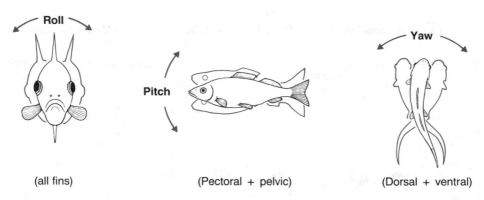

Fig. 21.24 How fish fins prevent instability

21.17 Amphibia

See also unit 2.4.

The frog (*Rana temporaria*).

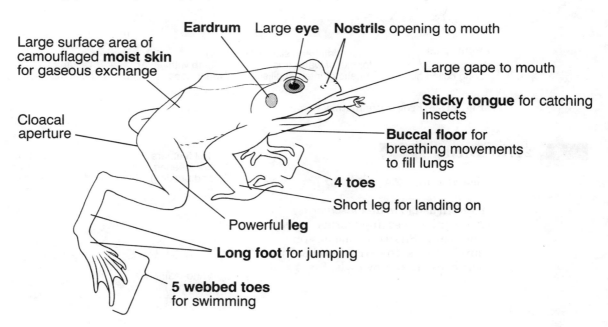

Fig. 21.25 Frog, external features

Adult: a carnivore, feeding on insects, worms in damp conditions and at night on land. Eaten by heron and other predators.

Tadpole: a herbivore, scraping algae off stones, plants; later becoming a scavenger on dead animals. Eaten by fish, water-beetle larvae, dragonfly nymphs and heron.

Life cycle

(*a*) In March, male frogs croak, inviting females into shallow water.

(*b*) Male grips female under arm-pits with swollen black 'nuptial pads' on thumbs.

(*c*) Female lays a few hundred *eggs;* male squirts *sperm* over them as they emerge in a continuous stream.

(*d*) Sperm must penetrate eggs to cause *fertilization* before albumen swells.

(*e*) Albumen gives egg *protection* from injury and predators; *camouflage* (by being transparent); and a *large surface area* for gaseous exchange.

(*f*) *Larvae* hatch (according to temperature) in about 10 days. In a *continuous* process of change (little happens overnight) larvae go through *three stages* (see Fig. 21.26).

(*g*) After about 90 days *young frogs* with stumpy tails hop onto land and start to catch insects with a sticky tongue. Hibernate in mud at bottom of ponds or in sheltered crevices, to avoid freezing each winter.

(*h*) Frogs are *adult* by their fourth season.

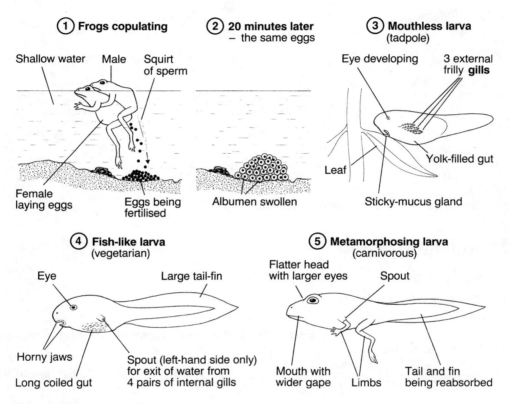

Fig. 21.26 Frog – life history

21.18 Birds

See also unit 2.4.

e.g. Thrush (*Turdus viscivorus*). An omnivore eating worms, insects and fruits. Snails it smashes on 'anvil stones' to extract and eat the soft body. Eaten by owls, foxes and cats.

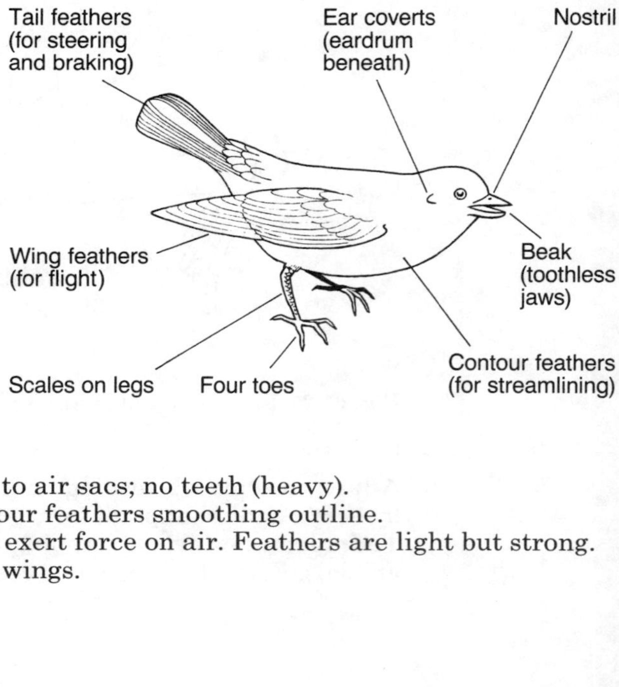

Fig. 21.27 External features of a bird

Adaptations for flight

(*a*) *Light bones* – some air-filled and linked to air sacs; no teeth (heavy).

(*b*) *Streamlined* – general body shape, contour feathers smoothing outline.

(*c*) *Feathered wings* – large surface area to exert force on air. Feathers are light but strong.

(*d*) *Large flight muscles* – on chest, flap the wings.

(e) *Large keeled sternum* – for attachment of flight muscles.
(f) *Special breathing system* and *large heart* – supply food and O_2 to flight muscles at a high rate. Flight requires a great deal of energy.
(g) *High body temperature* – ensures rapid respiration.

Flightless birds, e.g. ostrich, kiwi, lack one or more of these adaptations.

1 Flight

On wings – power
On tail – brakes

Quill

Shaft

2 Contour

On body – streamlining

3 Down

On body – insulation

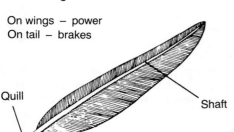
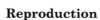

Fig. 21.28 Three kinds of feather

Reproduction

(a) Male birds of some species, e.g. robin, take up and defend *territories*.
(b) Courtship and display leads to *pairing* for the breeding season.
(c) One or both of the pair build nests in trees, holes or on the ground.
(d) Further display leads to *mating* and internal fertilization.
(e) *Eggs* are laid in nest singly, over a period of days, till clutch is complete.
(f) Eggs gain O_2 through shell from environment, and warmth from female's featherless brood-patches. She also turns the eggs daily, before *incubation* ('sitting').
(g) Chick *hatches* with help of egg-tooth (discarded after use).
(h) *Parental care* of young extends to removal of droppings, defence and feeding (instinctive behaviour induced by yellow gape of chick's mouth and chirruping, see unit 12.6).

21.19 Mammals

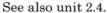

See also unit 2.4.
Mammal characteristics:

Reproduction

1 *Suckle* their young on milk from mammary glands.
2 *Viviparous* – give birth to young, not eggs (exceptions: echidna and platypus).

In the head

3 *Ear pinna* – external ear.
4 *Four kinds of teeth* – incisors, canines, premolars and molars.

Temperature regulation

5 *Endothermic* (homeothermic) – constant body temperature (birds also).
6 *Hair* – for insulation.
7 *Sweat glands* – for cooling.

Respiration

8 *Diaphragm* – muscular sheet separating heart and lungs from other organs.
9 *Erythrocytes* (red blood cells) – lack nuclei.

Most mammals are **placentals** – embryos are grown inside uterus (womb), nourished via the placenta. Others are **marsupials** – embryos are born early and grow mainly within a pouch, e.g. kangaroo. Two types only (echidna and platypus) lay eggs much like those of reptiles, but then give milk to the hatched young. These egg-layers are the **monotremes.**

21.20 Rabbit

The rabbit (*Oryctolagus cuniculus*): a herbivore eating a wide range of vegetation. Eaten by foxes, stoats, weasels.

Life history:

(a) Males pursue females in '*courtship*' chases'.
(b) *Mating* occurs mainly in January–June (but can be in any month).

(c) Female often digs a short (30–90 cm) 'stop' burrow away from warren at blind end of which she makes a *nest* of hay, straw and her own chest fur.
(d) *Gestation* of young takes 28–31 days; 3–7 blind young born.
(e) Female is often *mated* again within 12 hours of dropping litter.
(f) Young are *suckled* for about 21 days; fiercely protected from enemies.
(g) Young are *sexually mature* at 3–4 months; fully grown by 9 months.
(h) *Life expectancy* in wild about 1 year.

21.21 Bank Vole

The bank vole (*Clethrionomys glareolus*): an omnivore, eating mainly fruits, seeds, leaves and fungi; but also insects and worms in woods. Eaten by owls, kestrels, weasels and stoats. Active by day and night – moves along runways made amongst the vegetation or along burrows.

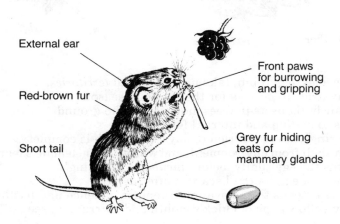

External ear

Red-brown fur

Short tail

Front paws for burrowing and gripping

Grey fur hiding teats of mammary glands

Fig. 21.29 Bank vole, external features

Reproduction: Breeding occurs from April to October. Females can breed at five weeks old giving birth to litters of four hairless blind young. Suckled on milk but weaned at two and a half weeks. But for their predators, vole populations could easily rise very fast.

22 BIOLOGY AS A SCIENCE

Biologists are scientists who study life. They are more than simply people who do nature-study, observing and describing. Biologists not only observe, they also experiment.

Living things are made of chemicals, and their chemistry is driven by energy. Such beings also have to obey the laws of physics. So a biologist must use the ideas of both physics and chemistry.

When biologists experiment, they make measurements. They use the same units of measurement as physicists and chemists the world over. These are called SI ('Systeme Internationale') units (see unit 22.3). Certain non-SI units are still in use, e.g. calories, litres (see Table 22.2 on p. 156).

Having made their measurements biologists often need to present them in a more meaningful way. They may show their results in a bar chart or graph; and they may also need to calculate averages, percentages and ratios. So they need some maths.

This unit will explain what is meant by the 'scientific method', the range of measurements used in biology and how to present them. It also summarizes the way in which certain common chemical reagents are used in biology. No matter which syllabus you follow you are expected to be familiar with how to use this information.

22.1 Scientific method

Biologists use the *scientific method* to discover new 'facts' —
1 They make **observations**, e.g. leaves are green; leaves make starch; greenfly feed on plants.
2 They have an idea or **hypothesis**.
This may try to explain what has been observed, or it may suggest a connection between two observations, e.g. leaves make starch *because* they are green.
Hypotheses come easily to mind when the following questions are asked about observations:
Who or what?   Why?   Where?   When?   How?
(Try out these questions on the three observations given in no. 1.)
3 They now **experiment** to see whether the idea is correct.
Every experiment must have a **minimum of two parts:** the test and the control.
The **test** is the part where a change is made to the biological material to test a hypothesis. The **control** is identical to the test except that the change is not made. If we were to test the hypothesis 'leaves in the light need carbon dioxide (CO_2) to make starch', leaves that were denied CO_2 would provide the test and leaves able to obtain CO_2 would provide the control.
To do this, certain leaves could be confined in flasks with a CO_2 absorbant; and others would also need to be put in flasks but with a CO_2 supply (see Fig. 5.3) to *compare* results. If the leaves *given* CO_2 did produce starch and those without CO_2 did not, one can be sure that confinement of the leaves within a flask was *not* responsible for the lack of starch. It must have been the *one* thing that was different that caused the lack of starch: lack of CO_2.
However, if both test *and control* leaves had produced no starch, it might have been possible to conclude that CO_2 is not important for starch-making in leaves – or at least the leaves of that particular plant (a strange result). Could the plant be diseased? It could have been that the leaves had been unable to make starch *in any case*. In other words, lack of CO_2 had not necessarily been responsible for the lack of starch in the test leaves.
Thus, the *control* is essential to an experiment to see whether
(*a*) the biological material is healthy;
(*b*) the apparatus or other conditions (also used in the test) may be playing a part in the results.
A further precaution in experiments is never to rely on results from one test and one control organism.
Experiments should always be done on **as large a number of organisms as possible** – as in an ideal class experiment. This makes sure that biological variation owing to sex, size, age, etc. is not having an effect on the results. For example, only 70% of carrot seed from a packet may germinate. If the conditions needed to germinate carrot seed were being investigated and only single seeds were used for each test and control, it could easily happen that one or more of the seeds chosen was a dead one. This would give misleading results. At least 20 or more seeds should be used in each test and control and the number of seeds germinating recorded as 0/20 or 14/20, etc.
4 The biologist now records his observations as **results**. These must be recorded accurately and honestly. Some very important discoveries in science have arisen by observations that were not expected and even 'unwanted'. For example, Sir Alexander Fleming's observation of a bacterial culture contaminated with a fungus led to the discovery of penicillin.
5 The biologist then **concludes** whether the results support the hypothesis. The conclusion is a new 'fact' – for a while at least.

22.2 Reporting your own experiments

You should use the following subheadings, in the order given here, to show that you are using the scientific method:
1 **Aim:** 'To discover ...' or 'To investigate ...' If you start off with 'Experiment to prove ...' it makes a mockery of what scientific investigation is all about. This start indicates to those assessing your work that the 'experiments' you did were actually *demonstrations* of something that you knew already – and some examiners may not give you much credit in consequence. So the Aim is a statement of the hypothesis you set out to test.
2 **Materials and method:** 'The materials were set up as shown in the diagram' could be the opening sentence. There follows a fully labelled diagram. Only information that the diagram does not explain needs to be added now, e.g. 'The plant used in the experiment had first been de-starched by keeping it in the dark for 48 hours' or, 'The length of the root was measured again after 24 hours'.
Note the use of the impersonal 'was measured'. Professional scientists prefer this to the personal statement '*I* measured'. A list of the materials used is unnecessary. Do not forget to emphasize which was the *test* and which the *control* part of the experiment.
3 **Results:** this should be *brief*, factual reporting of what happened – in both the test and the

control. Often the results can best be put into a table. Do *not* write an essay.

If, however, something unforeseen happened you may have to *describe* it and *explain* why it occurred in addition, e.g. 'Some of the germinating peas grew fungus and died'.

4 **Conclusion:** this should be a simple answer to the aim of the experiment. This is a plain statement, not an essay. Thus the aim might have been 'To find out if light is necessary for photosynthesis', and the experiment's conclusion is likely to have been 'Light is necessary for photosynthesis'.

22.3 Scientific Units of Measurement

Scientists use SI ('Systeme Internationale') units and have discarded the older units such as the foot (length) and pound (mass). A descriptive word called a prefix is put before the unit to show how large or small the measurement is. Thus a *kilo*metre (km) is a *thousand* metres and a *centi*metre (cm) is one *hundredth* of a metre – see Tables 22.1 and 22.2.

Table 22.1

Sizes						Unit					
10^{-9} (0.000 000 001) billionth	10^{-6} (0.000 001) millionth	10^{-3} (0.001) thousandth	10^{-2} (0.01) hundredth	10^{-1} (0.01) tenth	1 1.0	10^1 10	10^2 100	10^3 1000	10^6 100 000	10^9 1000 000 000
pre-fix	nano-	micro-	milli-	centi-	deci-	e.g. a metre	deca-	hecto-	kilo-	mega-	giga-
(and symbol)	(n)	(µ)	(m)	(c)	(d)		(da)	(h)	(k)	(M)	(G)

Thus, an object 1 metre long, i.e. 1 m, is also 100 cm = 1000 mm = 0.001 km long. The final 'm' in the abbreviation refers to the unit of length, the metre(m).

Mnemonic: Deca, Hecto, Kilo – go and fetch a pillow; deci, centi, milli – doesn't this sound silly.

Table 22.2 SI Units of measurement

Quantity	Unit	(prefix)	Useful examples Non-SI units also in use are in [ ]
Length	metre	(m)	Virus: 10 nm Bacterium: 1µm Cheek cell: 0.1 mm Large redwood tree 120 m
Area		(m^2)	Man's skin: 1.8 m² Lungs, at alveoli: 80 m² 1 hectare (ha) = 10 000 m²
Volume		(m^3)	Man's blood: 5-6 dm³ [1 dm³ = 1 litre]
Mass	kilogramme	(kg)	[1000 kg = 1 tonne(t) Large elephant: 5 t Large blue whale: 150 t]
Pressure	pascal	(Pa)	[Atmospheric pressure at STP = 101 kPa [or 760 mm of mercury]
Energy	joule	(J)	[4.2 J = 1 calorie] 4.2 J heats 1 cm³ water up to 1°C Teenage boy needs 12 MJ/day
Temperature	degree Celsius	(°C)	Human body: 36.8°C Boiling point of water: 100°C Refrigerator space: 4°C
Time	second	(s)	[1 minute (min): 60 s 1 hour (hr): 3600 s 1 day(d): 86 400 s 1 year(a) $\simeq$ 3.158 × 10⁷ s]

Solutions

A solution is a **solute** dissolved in a **solvent**, e.g. sugar dissolved in water. The *concentration* of the solution can be measured in g/dm³. It can also be expressed in moles.

A *molar* solution is the molecular mass of the substance dissolved in 1 dm³ of solution. Example: the atomic mass of carbon (C) is 12, of hydrogen (H) is 1 and of oxygen (O) is 16. The formula of glucose is $C_6H_{12}O_6$

The molecular mass of glucose is thus
$$12 \times 6 = 72 \quad (C_6)$$
$$1 \times 12 = 12 \quad (H_{12})$$
$$16 \times 6 = \underline{96} \quad (O_6)$$

The molecular mass of glucose is therefore   180 $(C_5H_{12}O_6)$

A molar solution of glucose is thus made by adding 180 g of glucose to a litre measuring flask and topping it up to the 1 dm³ mark with distilled water. Glucose will remain as $C_6H_{12}O_6$ molecules, in the solution.

Certain other substances will, however, split up into particles that carry electrical charges, called **ions**. Common salt (NaCl) solution contains Na^+ (sodium ions) and Cl^- (chloride ions).

The **pH scale** is a measurement of the concentration of hydrogen ions (H^+) in a solution. Certain *indicators* show whether solutions are acid (pH 1–7), neutral (exactly pH 7) or alkaline pH 7–14) by changing colour according to their pH. (See Table 22.3.)

Table 22.3

Indicator solutions	Very acid Acid pH 1 ← ———————		Neutral pH 7	Alkaline —————————	Very alkaline → pH 14
Litmus		← ——————— red	purple	blue ———————————	→
Phenol phthalein	← ————— colourless —————————			pH 8.4 red ——————————	→
BDH Universal indicator	pH 4 pH 6 red orange	pH 6.5 yellow	pH 7.0 green		pH 9.0 pH 11 blue purple

Solutions are all-important both inside cells and also outside them:

1 The 'strength' (concentration) of the solution inside a cell determines its osmotic (= water) potential, i.e. its ability to take up water – or to lose it – to a neighbouring solution. Plants absorb water in this way (see unit 7.3).
2 The pH of cells, or in the gut, is very important if enzymes are to work properly (see unit 1.5). If enzymes fail, cells die.
3 Certain ions are used by nerve cells to generate electrical impulses – their method of passing 'messages' (see unit 12.3).

22.4 Elements, Compounds and Mixtures

Elements are substances which under normal circumstances cannot be changed, e.g. carbon (C), oxygen (O), hydrogen (H) and nitrogen (N). They do not even change their identity when they combine with each other. Elements are made of minute particles of identical mass called atoms, e.g. ^{12}C has an atomic mass of 12.

Certain elements have two or more different types of atom called *isotopes*, each with a different mass, e.g. ^{12}C and ^{14}C. Some isotopes are *stable*, e.g. ^{12}C, but others are *radioactive*, e.g. ^{14}C, and give out radiations that will expose a photographic plate. Normal oxygen ^{16}O has a heavier isotope ^{18}O which is also stable but this too can be detected, using an instrument called a mass spectrometer. Both kinds of isotope can be used as *tracers* to follow chemical reactions in organisms (see units 5.1 and 7.8).

Compounds are combinations of elements in fixed proportions. The proportions can be written down as a formula which shows the numbers of atoms of each element present, e.g. CO_2 represents one atom of carbon combined with two atoms of oxygen. A single particle of a compound is a **molecule** (see Fig. 22.1).

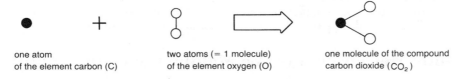

one atom
of the element carbon (C)

two atoms (= 1 molecule)
of the element oxygen (O)

one molecule of the compound
carbon dioxide (CO_2)

Fig. 22.1 The carbon dioxide molecule

Inorganic molecules are small, simple ones like CO_2, H_2O, and $NaNO_3$. They are plentiful outside organisms in non-living matter.

Organic molecules are usually large and complex in structure. They usually contain two or more carbon atoms, e.g. $C_6H_{12}O_6$ (glucose). Organisms make organic molecules, examples being carbohydrates, fats (lipids) and proteins (see units 4.1 and 4.4).

Living things are made up of *mixtures* of water, inorganic and organic molecules. There is no set proportion of elements and compounds in a mixture, so no formula can be written for it.

22.5 Energy

The sun radiates waves of energy to the earth. Some waves have a short wavelength, e.g. X-rays; others have a long wavelength, e.g. radio waves. This range of wavelengths is known as the electromagnetic spectrum. Only the shorter waves have any known biological importance – see Table 22.4 on p.158.

The *light* energy trapped by plants is stored as *chemical* energy in bonds of organic compounds, e.g. glucose. As glucose is broken down during respiration, energy is released for movement, so providing *mechanical* energy but much is wasted as *heat*. Thus, energy that came into living things as light may be converted into a variety of other forms of energy (see unit 9.13).

Too much heat energy may kill organisms. It may be lost by *conduction, convection, radiation* and by *evaporation* of water. However, too little heat may also kill, so it may be kept in by *insulation* (see unit 10.7).

Table 22.4

Wavelength	700 nm		360 nm
Type of wave	Infa-red	Visible light Red Orange Yellow Green Blue Violet	Ionizing radiations Ultra-violet X-rays γ-rays
Effect	Heat (see unit 10.6)	trapped in photosynthesis (see unit 5.1)	can cause mutations (see unit 17.15)

22.6 Surface Area to Volume Ratio

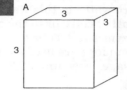

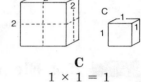

Fig. 22.2 Surface area in proportion to volume

As an organism gets larger (in volume) its surface area does not increase in proportion (see Fig. 22.2):

	A	**B**	**C**
Surface:	$3 \times 3 = 9$ cm^2 (each face)	$2 \times 2 = 4$	$1 \times 1 = 1$
	$6 \times 9 = 54$ cm^2 (six faces)	$6 \times 4 = 24$	$6 \times 1 = 6$
Volume	$3 \times 3 \times 3 = 27$ cm^3	$2 \times 2 \times 2 = 8$ cm^3	$1 \times 1 \times 1 = 1$ cm^3
Ratio = $\dfrac{\textbf{Surface}}{\textbf{Volume}}$	$\dfrac{54}{27} = \textbf{2}$	$\dfrac{24}{8} = \textbf{3}$	$\dfrac{6}{1} = \textbf{6}$

Let us assume that A, B and C are living. In C a volume of one cubic centimetre (cm^3) has 6 cm^2 through which oxygen or food enter, and through which CO_2 and other wastes can leave. But in B each cm^3 has only 3 cm^2 through which these functions can occur. This makes the movement of these materials more difficult. It is not surprising therefore that organs concerned with **absorption** (guts, lungs, leaves, roots) or **excretion** (lungs, kidneys) have large surface areas.

Surface area is equally important where **heat loss** is concerned. Small animals, e.g. shrews, with their large surface area to volume ratio, lose heat very easily. They eat a lot of food to provide heat to make up for this loss. Elephants on the other hand, being large, have the problem of keeping cool since they have a small surface area for their bulk. They resort to bathing; but also use their large ears to radiate heat (see unit 10.7).

The surface area of food is increased by chewing. In Fig. 22.2, if B were a 2 cm cube of food and it were chopped into eight separate cm cubes, the total surface area would increase from 8 cm^2 to 48 cm^2. This would give digestive enzymes more area to work on.

22.7 Handling Measurements and Making them Meaningful

A class of 20 pupils gave the following information shown in a Table 22.5.

Table 22.5

BOYS			GIRLS	
Height/cm	Blood Group		Height/cm	Blood Group
124	AB		132	O
132	B		129	A
144	O		126	O
129	B		143	A
133	O		141	A
139	A		136	A
138	O		131	O
147	A		121	AB
136	B		134	O
138	B		137	O
——— (total)			——— (total)	

Combined total of boys and girls _____ *cm*

Height measurements provide a *continuous* range of readings, i.e. each one can differ from the next one by only 1 cm. These results can be expressed as a diagram called a **histogram** (see Fig. 22.3a).

However, the blood groups are *discontinuous*, i.e. there are only four types. These can be expressed in a similar diagram called a **bar chart** (= bar graph). The columns in Fig. 22.3(b) are separated from one another.

(a) **Histogram** of heights of pupils

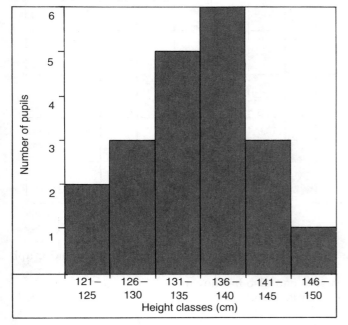

(b) **Bar chart** of blood groups of pupils

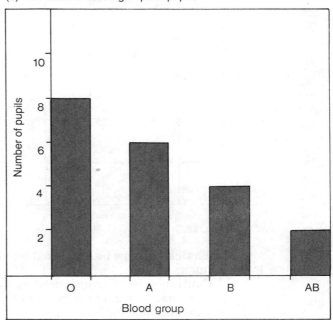

Fig. 22.3

Note that it is very important to put in the measurements you are using: *numbers of individuals* on the vertical axis (y axis) and *measurement* or characteristic on the horizontal axis (x axis). A *title* must be added.

For practice, draw in the histogram for height of *boys*; and the bar chart for blood group of *girls* in Fig. 22.4.

Histogram of the heights of boys

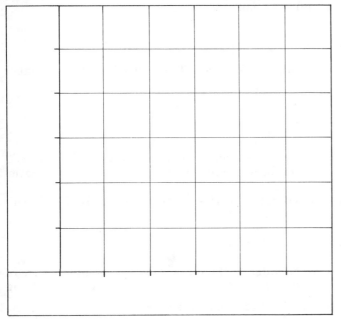

Title: _ (girls)

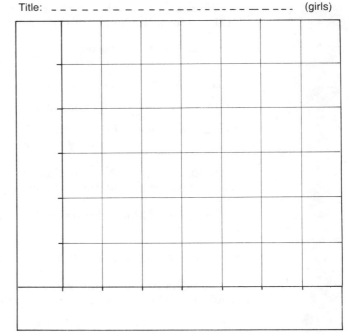

Fig. 22.4

To do this, start by entering a tick on a chart such as that shown below, for each height class you decide upon. The first three boys in the list have been ticked for you already. Do something similar for the girls' blood groups (Fig. 22.5).

Height classes for boys (cm)

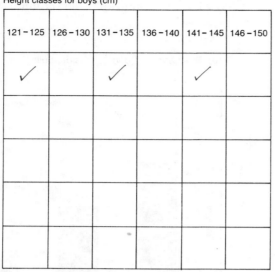

121 – 125	126 – 130	131 – 135	136 – 140	141 – 145	146 – 150
✓		✓		✓	

Fill in a tick for each boy

Blood groups for girls

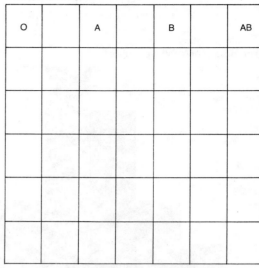

O		A		B		AB

Fill in a tick for each girl

Fig. 22.5

Each tick can now be converted into one of the blocks that make up the histogram or bar graph columns.

Continuous or discontinuous measurements can also be shown as **pie-charts,** e.g. blood groups of the class (Fig. 22.6.) Try making one for boys only.

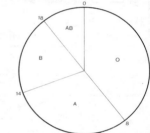

Fig. 22.6 Pie-chart of blood groups in a class

Method:

The number of O blood group in the class is 8. There are 20 in the class. A circle has 360°. So O blood group will have $\frac{8}{20}$ × 360 = 144° of the 360° pie.
Using a protractor, the appropriate 144° line can be drawn in.

The **range** of heights in the class is given by taking away the height of the shortest person (121 cm) from the height of the tallest (147 cm) = 26 cm.

Calculate the range for boys: _____
and for girls: _____

The **average** (= mean) height is the sum of all the heights (2269 cm) divided by the number of pupils (20): $\frac{2269}{20}$ = 113.45 cm
Calculate the mean height of the boys: _____
and of the girls: _____

Histograms, bar charts and pie charts all show measurements in a visual way, which should make them easily understood. You may, however, be asked to use the information mathematically.

The **ratio** of blood group O to blood group AB in the class is 8:2 or, more simply, 4:1 (four to one).

Calculate the ratio of A to B blood groups in the girls of the class: _____

The **percentage** of pupils that are 140 cm or taller in the class refers to the number there would be if the class were 100 strong ('cent' means a hundred'). It is calculated by multiplying the ratio by 100. Thus, 4 pupils are over 140 cm tall and the class is 20 strong. The ratio is $\frac{4}{20}$. The percentage is $\frac{4}{20}$ × 100 = 20%.

You may also be asked to select information more carefully, e.g. what percentage of the class is boys with blood group A: $\frac{2}{20} \times 100 = 10\%$.

Now calculate the percentage for the girls: _____

A **graph** is another visual means of presenting data (scientific measurements). One of its values is to allow predictions to be made from it. Data on the mass of American boys could be presented as a graph (see Fig. 22.7).

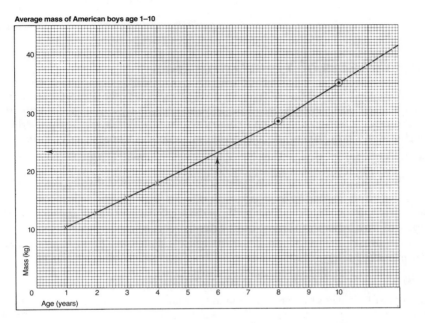

Average mass of American boys age 1–10

Data	
Age/yrs	**Average body mass/kg**
1	10
2	12.5
3	15.0
4	17.5
8	28.0
10	34.5

Fig. 22.7

Note that a *scale* must be chosen for the measurement (in this case 10 kg = 10 units on the *y* axis; and 1 yr = 10 units on the *x* axis). This scale must be a sensible one if the graph is to be fitted onto the graph paper and will therefore depend on the range of measurements given. This range is 34.5–10.0 = 24.5 kg and 10−1 = 9 years. However, it is usual to start at 0 for both *x* and *y* axes and to allow a bit extra.

The data are then plotted by putting either a cross × or a ringed dot ⊙ exactly where readings for age and average body mass intersect. These dots or crosses are then joined up.

From the graph, *predictions* can be made:

What is the average mass of American boys at age 6? (see above)

What is the average mass of American boys at age 12 likely to be?

Here the line can be projected (see above) and a reading taken.

Read from the graph what age you would expect a 14 kg boy to be.

22.8 Drawings

It is not easy for some people to record their observations of organisms as drawings. This may help you:

1 Measure the object and then scale it up before you draw on paper (Fig. 22.8).

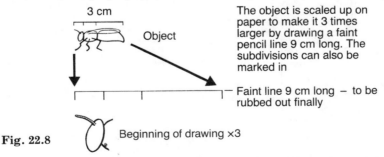

3 cm

Object

The object is scaled up on paper to make it 3 times larger by drawing a faint pencil line 9 cm long. The subdivisions can also be marked in

Faint line 9 cm long − to be rubbed out finally

Fig. 22.8 Beginning of drawing ×3

These marks help you to draw the object in correct proportion.

2 Once drawn in pencil, the drawing should be given a *title* and a *scale* added (in this case '×3').

3 *Labels* should be added, as required, by ruling lines in biro or pencil, and adding names (see Fig. 22.9).

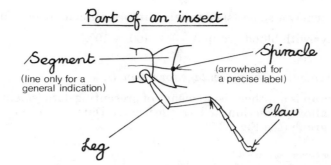

Fig. 22.9

22.9 Ideas for Experiments of Your Own

Below are some ideas for you to choose from in designing your own experiments as part of your Coursework Assessment. Once you have chosen, read units 22.1 and 22.2, and perhaps 22.8. The suggestions below have a reference to the relevant unit in this book which will give you the background to the idea and often some suggestions as to the apparatus you may wish to use.

How does the vitamin C content of food vary? (ref: unit 4.5)

(a) Measure the amount of vitamin C in fresh new potatoes and old potatoes; fried potatoes, boiled potatoes and potato water.

(b) Does it matter *how long* the potatoes were stored, or cooked?

(c) Do conditions of light/dark and heat/cold affect (b)?

(d) Measure the vitamin content of cabbage, lettuce or milk (if you can get it fresh from the cow) – bearing in mind (b) and (c)

What conditions affect the rate of photosynthesis? (ref: unit 5.4)

(a) Measure the rate of photosynthesis in *Elodea* (or other submerged water-weeds). Apparatus B is more accurate – with apparatus A, are all bubbles the same size? Vary (i) the light intensity, or (ii) the CO_2 level, or (iii) the temperature – keeping the other conditions constant.

(b) What effect do (i) muddy water (*how* muddy?), (ii) green water (many unicell algae in it), (iii) depth of water (many water baths in a row) have on the rate of photosynthesis?

Does the rate of respiration of a pond snail change as the pond gets warmer? (ref: unit 5.6)

Measure the rate of respiration of pond snails at different temperatures. Blow through some stock *red* hydrogen carbonate indicator solution in a test tube with a straw. When it just turns *pale yellow*, add a rubber bung. This is your colour-reference tube.

1 Warm (or cool) red indicator solution in one stoppered tube; and a snail in a second tube which is left open, in a water bath for a few minutes.

2 Then place the snail in the red indicator tube and re-stopper. Time how long it takes for the red indicator to turn to the colour of the colour-reference tube.

Repeat for other temperatures using the same colour-reference tube, same snail but fresh indicator solution.

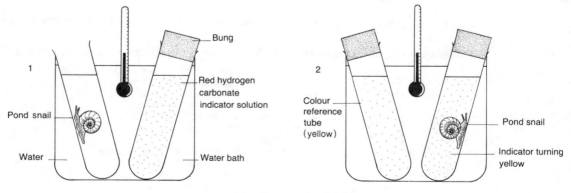

Fig. 22.10

What happens to the heart beat of *Daphnia* as the pond gets cooler? (ref: unit 21.6)

Measure the heart beat rate of *Daphnia* at different temperatures. With a dropper put a *Daphnia* onto a cavity slide with some water. Place the slide on top of a petri dish containing water at a known temperature. Observe under a microscope.

How much does a supply of mineral salts to seedlings affect their growth? (ref: unit 5.8)

Measure the gain in mass of batches of seed (e.g. 100 mustard seeds) grown in different mineral salt solutions. Other growing conditions must be the same. Grow batches on weighed blotting paper dipping into the different mineral salt solutions. Harvest and dry the batches in an oven at 100°C after a few weeks' growth; reweigh.

(*a*) Use solutions each lacking a different element, e.g. no nitrogen.

(*b*) Use a full culture solution to which is added extra quantities of an element, e.g. × 1, × 2, × 4 the amounts of nitrogen.

How do the quantities of starch, of hydrogen ions and of salivary amylase affect the rate of digestion of the starch? (ref: unit 6.3)

Measure the rate of digestion by salivary amylase under constant conditions:

(*a*) when the *quantity of enzyme* is altered, e.g. tubes with 1, 2, 3 and 4 cm^3 of enzyme solution added to the 5 cm^3 of starch solution in each;

(*b*) when the *quantity of starch* is altered, e.g. tubes with 5, 10, 15 and 20 cm^3 of starch, to each of which is added 1 cm^3 of enzyme;

(*c*) when *the pH* is changed, e.g. by adding acid or alkali (see unit 6.3, Example 2).

How strong is the solution inside rhubarb cells? (ref: unit 7.4)

(*a*) Measure the concentration of the solution inside rhubarb epidermis cells. Strip off red epidermis about 0.5 cm wide from rhubarb stalks. Cut off 1 cm lengths from the strip with scissors so that they drop directly into drops of sugar solution of different concentration, placed on separate slides. Add a coverslip to each and observe under the microscope after 2 minutes. Record how many cells out of 20 are plasmolysed or turgid. How does this give you the answer? Do other solutions of the same molarity, e.g. salt or glucose, give the same result? Do cells from wilted rhubarb give a different result from cells of healthy rhubarb?

What makes the pores of a leaf (stomata) open and close? (ref: unit 7.5)

Discover what factors cause opening or closing of stomata in a leaf. Use the nail varnish technique, having left the leaf in one of the following conditions for 5 minutes: light or dark; warm or cool; with CO_2 (see Fig. 5.3) or without it. While testing one pair of conditions, keep the other conditions, e.g. temperature and CO_2, as constant as possible. Do not use hairy leaves; shiny ones are best.

Do small animals such as beetles and woodlice respire faster when it is warm? (ref: unit 9.5)

Measure the rate of respiration of a small animal at different temperatures. Allow it to gain the temperature of the water-bath with the syringe nozzle pulled out for 3 minutes. Return the nozzle to the bung and suck an ink drop into the capillary tube. Measure how far the drop moves in 5 minutes. Do the animals need to be kept still by a wad of cotton wool?

(*a*) Use locust; earthworm; woodlouse, etc (any one).

(*b*) Do different species of animal differ in their rate of respiration (on a *weight for weight* basis) at a particular warm temperature?

How does ethanol (alcohol) affect a person's coordination? (ref: unit 12.9)

Measure the effect of alcohol on the nervous system. Test the rate of reaction of your parents or other members of the family before and after they have been to the pub. Record what they have drunk and over what time period.

(*a*) Place your forearm on a table and your parent's forearm on a chair. Drop the metre ruler vertically between your parent's forefinger and thumb. Ask your parent to catch the ruler as it falls. Measure the distance the ruler falls.

(*b*) Play a video game, e.g. 'squash', 'tennis' – anything that requires anticipation and motor skills. Record the scores.

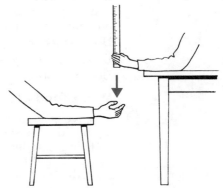

Relate (a) distance fallen, (b) video scores, to amount of alcohol consumed. Do tiredness, time of day or amount of food eaten by your parent affect your experiments?

Fig. 22.11

How do some weeds defeat the gardener trying to get rid of them? (ref: unit 14.2)

How large a piece of rhizome (underground stem) is necessary to form a new plant? Use weed species such as ground elder (*Aegopodium*), bindweed (*Convolvulus*) or couch grass (*Agropyron*). Cut the cleaned rhizome into lengths (e.g. 2.5 cm and 10 cm) and weigh them. Plant in damp potting fibre in a warm place. Count those that show a leaf above ground. Unearth all of them to see whether length or mass or presence of a node is the deciding factor. Anything else? Do winter, spring and summer rhizomes behave differently?

What makes pollen germinate? (ref: units 14.7 and 14.8)

(*a*) Measure the percentage germination of pollen in water and various concentrations of sugar solution. Place three layers of blotting paper in a petri dish. Add a flat strip of Visking tubing. Pour sugar solution (e.g. 5, 10 or 15%) onto the Visking until the blotting paper is just flooded. Dust ripe pollen from ripe stamens onto the Visking in each of the dishes; add a lid and leave for 24 hours. Lift the Visking onto a slide with forceps and count the percentage of germinated grains under a microscope.

(*b*) Does temperature or light affect germination?

(*c*) Do all flower species germinate best at the same sugar concentration?

(*d*) Do pollens from wind-pollinated and insect-pollinated flowers die off after a time and at the same rate? Collect pollen from two species of flower in separate tubes. Dust pollen at different times (e.g. collected fresh and after 1, 4 and 8 days' storage) onto Visking as in method (*a*).

Do seedlings lose weight during their first week of growth, just as new-born babies do? (ref: unit 16.6)

Measure the change in mass of seeds (e.g. broad beans) as they grow into young plants. Measure both the live (wet) mass and dead (dry) mass at intervals; make a graph of the results; explain what three processes account for the changes in mass.
Example
Weigh five batches of 10 seeds, A–E, and plant 5 cm down in damp potting fibre. Using batch A, reweigh every 2 days and replace in the fibre. Do the last weighing when most seeds have radicles peeping through the testa. Then obtain batch A's dry mass after drying the batch in an oven for 24 hours at 100°C.

Measure wet and dry mass of other batches as follows –

B: 4 days after last weighing of A; C when half the seedlings are just peeping through the fibre; D and E when the batch has been above ground in sunlit conditions for 1 and 2 weeks respectively.

How fast does a population of duckweed grow? (ref: units 16.7 and 19.3)

(*a*) Measure the rate of increase in population of duckweed (*Lemna*). Using a paintbrush, transfer 10 healthy (green) duckweed plants from a pond to a 250 cm^3 beaker containing 200 cm^3 of 'full culture' solution. Cover with pin-holed 'clingfilm' and put in a warm, light place to grow. Count the number of plants every 4 days by transferring them to a fresh beakerful of culture solution.

(*b*) Compare the rise in population of duckweed placed in beakers with different quantities of solution (e.g. 25, 50, 100 and 200 cm^3) Do *not* transfer them to fresh solutions (compare (*a*) above).

What precisely does the water-louse Asellus eat? (ref: unit 19.5)

Discover the food preferences of *Asellus* and their role in decomposition of leaves. Cut measured squares of freshly fallen oak leaves. Soak five in pond water and five in muddy pond water for a week. Leave five dry. Add the three batches of five leaf squares to different jam jars half full of pond water containing five *Asellus*. After a number of days remove the squares and trace their outlines on graph paper.

Do *Asellus* prefer oak to any other species of fallen leaf?

Do the leaves decay better on their own without *Asellus* in the jars?

Does the pH change (or the appearance of the water) during the process?

What food attracts scavengers in a pond? (ref: Fig. 19.9)

Discover the food preferences of pond scavengers. Tie string round the necks of jam jars and anchor the loose ends with stakes to the bank. Put various baits in the jars, e.g. fresh meat or rotting meat; dead snail or rotting snail; fresh chopped or dead chopped leaves, etc. Sink these in water and haul in for study at intervals, e.g. every 20 minutes or every hour. Count and describe the kinds of organism. This is best done in summer.

Does the position around the pond affect results?

Does the depth of water or type of bottom (e.g. mud or gravel) affect results?

Does it matter how you set up a pit-fall trap? (ref: Fig. 19.12)

Discover the differences in pit-fall trap catches when set up in different ways. Set up a number as shown in Fig. 19.11; another number, dry; another number with 5% formalin; another number with baits, e.g. a small piece of meat. Leave overnight, collect in the morning and observe carefully *everything* in the jar.

Do the following affect your results: the vegetation around each jar; the amount of leaf litter; the weather beforehand? This is best done in the autumn.

How do the small animals in the soil react to rain – or the lack of it? (ref: Fig. 19.14)

Discover the effect of rain on the distribution of invertebrates in the soil. Take horizontal layers of soil (e.g. 5 cm thick) from the same area and extract the invertebrates by Tullgren funnel. Do this after heavy rain, a few dry days later and after a dry spell. This is best done in September.

HINTS FOR CANDIDATES TAKING BIOLOGY EXAMINATIONS

Showing the Examiner What You Know, Understand and Could Do (by Experiment)

Students should not be entered for an examination that is beyond their ability. Success in examinations for which you have been entered (which assumes that you *do* have the ability) lies in good 'examination technique'. Your teacher will usually advise you on the type of examination you will sit by showing you past question papers (see also specimen questions on pp.169–207). But certain principles of technique apply to all methods of examination:

1 **Come fully equipped** with pen, pencil, rubber, ruler and coloured pens or pencils.

2 **Read the exam instructions carefully** — do not leave out compulsory questions.

3 **Plan your time for answering** according to the marks allocated. If you are given 40 minutes to complete 50 multiple choice items, you can calculate that you have 48 seconds per item. More usefully, you can work out that you should at the very least have reached Question 25 after 20 minutes in the exam room.

4 **Do the maximum number of questions.** Usually, modern exams allow you plenty of time to complete all the questions. But always check your answers right through for any that you may have missed.

   Where you have left an answer-space blank you can be certain of one thing: a *blank* scores *no* marks. So look again at the question and write *something*. That 'something' has a better chance of scoring you marks than a blank. Think positively!

5 **Choose the right questions to do.** There is little or no choice in modern exams. But where choice exists — usually in papers for those attempting A and B grades — look carefully at the mark allocation shown in brackets on the question paper. Each mark usually rewards an idea or fact that you can put down. By adding up the number of ideas and facts you can put down for each section of the question, you can gain some idea of the number of marks you would score. By comparing your scores for each of the questions, you can make the best choice.

6 **Understand what the question asks.** Never twist the examiner's words into a meaning that was not intended. It is no use answering a completely different question from that written on the examination paper. If someone asks you how to mend a bicycle puncture and you reply with an excellent description of how to raise the saddle, you have not answered their question, nor have you given them any help! In an examination, mis-information of this kind earns you no marks. You may know the correct answer all along. But if you fail to show the examiner that you know, how can you succeed?

7 **Plan before writing** your essay, paragraph and experimental answers. Organize key words into a *logical order* or pattern. This is particularly important when you have been asked to design an experiment. Use *short*, clear sentences, each one explaining a single step in the procedure. See the section below and unit 22.2

8 **Use large labelled diagrams** in your answers if they make your answer clearer. Descriptions

of experiments are almost always clearer, and certainly much shorter, when diagrams are used. Diagrams save words.

9 **Set out your work neatly.** An examiner is human. If your written answers are neatly set out he is much more likely to give you the benefit of the doubt where your answers are not entirely clear.

10 **Keep a cool head.** You can only do this by getting plenty of sleep and some exercise over the examination period. You will reason better if you do *not* stay up all night revising.

Tackling Various Types of Question

1 Multiple choice questions

These are sometimes called fixed response or objective questions. At first sight these questions seem to be comparatively easy because answering them is simply a matter of choosing one correct answer from the possible answers given. However, the questions are designed to test how well you understand specific topics and students do not always obtain as high a mark as they expected. But providing you know or can work out each answer (see below), the multiple choice questions in an examination should not be troublesome.

If four choices of answer are offered, usually two are very obviously wrong. You now have a 50% chance of being right even if you don't know the answer. Don't leave the odds at 25% by a blind guess.

Suppose the question requires you to *reason* from facts you should know. Say the question is 'Which gas(es) are produced by a green plant's leaves in the dark?' and the answer choices are: (a) CO_2, (b) N_2, (c) CO_2 and O_2, (d) O_2. From these (b) can be eliminated because nitrogen gas is neither used nor produced by green plants on their own. 'Leaves' readily suggest photosynthesis, a by-product of which is oxygen. But *light* is needed for photosynthesis and the question states that the leaves are in the *dark*. So (c) and (d) must be wrong because both include oxygen. That leaves (a) as the answer. There are also other types of 'choice' questions which are more testing.

2 Essay and paragraph questions

Most essay questions today are 'structured' into sections which require paragraph answers. The principles for writing essays or paragraphs are the same. The examiner is looking for a number of points that you should be remembering as key words — just how many is often suggested by the mark allocation. On rough paper write down the key words and join these by lines into a pattern-diagram where necessary. Number the key words according to the order in which you are going to use the facts in your answer. In this way your facts will be presented logically; and nothing will be left out. If *examples* make your answer clearer, use them.

Take the following example of a structured question:

'(a) Why are enzymes frequently referred to as "biological catalysts"?   (4)
(b) What are the effects of changing (i) pH, (ii) temperature upon the rate of action of any **named** enzyme?'   (7)
(there followed a third section to complete the question.)

The way to go about planning your answers is illustrated below:

Notice that this student used the key words from the question to build up this pattern-diagram

From this pattern-diagram, done in a minute or two, might come a written answer like:

(a) Catalysts are substances which in small amounts can greatly increase the rate of certain chemical reactions. Catalysts remain unchanged at the end of the reaction. Enzymes, unlike catalysts used in chemical works, are proteins. They control the rate of reactions in living things e.g. in respiration and digestion.

(i) Pepsin digests proteins in the stomach where conditions are acid. It will not do so if conditions are alkaline.

(ii) Pepsin works best at body temperature. If it is boiled it is destroyed and stops working. If it is cooled by ice it will also stop working but it is not destroyed.

'State' and 'explain' questions: 'state' or 'list' means put down as simple facts — nothing else. 'Name' is a similar instruction: no explanations are required.

'Explain' requires not only the facts or principles but also the reasons behind them. When you are thinking out the answer to an 'explain' question, ask yourself 'which?', 'what?', 'where'?, 'why?' and 'when?' about the subject. These questions will help you to avoid leaving out information that you know. You *must*, however, only give the information that is asked for — for example 'which?' and 'when?' may be irrelevant (unnecessary) in a particular question.

'Calculate' usually means not only give the answer but *show your working*.

'Deduce' usually means reason out an answer and *state your reasoning*.

Consider these questions:

1 'State three features commonly shown by animals at their respiratory surfaces.'

The answer could be: 'Large surface area; wet surface; often associated with a blood system' — to give the bare essentials. No *reasons* are required; and only three lines were allocated for the answer.

2 (a) *Name two* enzymes, secreted by mammals, which digest carbohydrates.
(i) ....
(ii) ....(2)

(b) Select *one* of these enzymes, *name* the substance which it digests and state the product or products formed.
Name of enzyme..
(i) Substrate...
(ii) Product or products..(2)

When reading through questions you have decided to answer, **underline vital words.**

Note that the examiners have been particularly helpful here by putting in italics vital words that *you* would have underlined. However, as you read through it would be worth underlining 'carbohydrates'. The remaining words that you would have underlined are already emphasized for you after the numerals (i) and (ii).

The answers could be:

(a) (i) salivary amylase (= ptyalin)
(ii) maltase

(b) salivary amylase   **OR**   maltase
(i) starch   maltose
(ii) maltose   glucose

'Compare' and 'contrast' questions: 'contrast' means pick out the *differences* between. If you are asked to do this you must use such words as 'Whereas ...' and 'however ...' It is not sufficient to give two *separate* accounts of the two organisms or processes to be contrasted.

'Compare' means pick out not only *differences* (contrasts) but also *similarities*. Thus your answer will include not only 'whereas ...' and 'however ...' statements but also 'both ...'

In planning such answers it is vital to write down on rough paper, in three columns, the features to be compared or contrasted and, alongside, the comparison you have made mentally.

Feature, or characteristic	Organism, or process A	Organism, or process B
1 ... 2 ... etc	Differences (i.e. *contrasts*)	
1 ... 2 ... etc.	Similarities	

} *Comparison*

Such questions are usually only found in papers taken by those of higher ability.

3 Graph, diagram and experiment questions

Graphs: If you are asked to put information onto a graph, it is vital that on both axes you state the relevant *units*, e.g. 'g' or 'cm³/h' or 'numbers of live insects'. Usually the title of the graph is supplied by the question — but sometimes it is important that *you* should provide it. All plots must be precise and ringed. You will avoid wrong plots by using a ruler to lead your eye to the precise spot. Join each plot with a *straight* line to the next one.

Diagrams:

1 *Draw in pencil* — in case you need to use an eraser.
2 *Draw large* — for clarity and easy labelling; then put down your pencil.
3 *Rule your labelling lines* in biro, avoiding crosses. Biro does not smudge against the ruler. Neither can the straight biro labelling lines be confused with being part of the detail of the drawing (which is in pencil).
4 *Label* in ink or biro neatly and add a *title*.
   If you follow this drill *in sequence* you will save time. And time is often marks!

Experiments

Experiments must be written up in a logical order under subheadings. The account usually includes a diagram which *saves* words. Do not duplicate the information in a diagram by also giving a *written* account of what it shows. Only write what the diagram does *not* say.

Questions asking you to *describe* experiments are usually of two types:
1 **Coursework assessment** of an experiment that you have carried out. Below is a reminder of the way that such an account should be set out.
2 **Examination questions** which usually restrict themselves to asking you to *design* an experiment. These answers require only the 'Materials and Method' part of the sequence outlined below (see also unit 22.2).

Aim: you should start with 'To discover...' or 'To investigate...' — never 'Experiment to prove ...' State the hypothesis (idea) you are testing in your experiment.

Materials and method: start with 'The materials were set up as shown in the diagram below'. Now draw a fully labelled diagram of your experiment. The 'test' and the 'control' parts must be clearly indicated. Finally, give any *extra* information not shown by the diagram. For example, 'The seeds were reweighed every two days. On day 10 they were crushed and tested with Benedict's solution and with iodine'.

Results: write a plain statement of what happened in both test *and* control — no discussion. Record your results in a table if at all possible. Discussion, if needed, can follow.

Conclusion: end with a simple answer to the question posed in the Aim. For example, the aim 'To discover whether chlorophyll is necessary for photosynthesis' is likely to be answered by the conclusion 'chlorophyll is necessary for photosynthesis' — and nothing more.

Too often students score poor marks by giving rambling accounts which leave out important details. Accounts which are organized into a logical sequence always score better.

4 Relevance in answers

Sadly, a large number of reasonably knowledgeable students do not do themselves justice by writing irrelevant answers. Sheer length of an answer will not gain any marks. It is only the key facts and principles that the examiner is looking for, *whatever* the length of the answer. So do not 'pad out' your answers.

The length of answer required is often suggested either by the marks awarded to it in the mark scheme (usually stated alongside the question), or by the space allocated to it on an answer sheet. If your answer is about to be either much shorter or much longer than these two indicators suggest, think again. Re-read the question — and your underlining of the important words in it.

The following questions are all specimen questions provided by Examining Groups to give a taste of what GCSE and Scottish Standard Grade exams are like. They have been carefully selected

(a) to give as wide a range of subject matter as possible

(b) to illustrate the range of question types

(c) to illustrate the way in which examiners mark answers – see answer section, pp. 207–213

The answers given are the author's own, to help your understanding, but they are heavily based on the answers provided by the Examining Groups with their questions – and particularly their mark allocations. The answers given are not necessarily the only ones that would gain marks.

The purpose in providing both questions and answers is to give you some practice. Answer the questions on paper after you have revised each topic. Only then check your answers against the answers given in the book. If you find any questions difficult, refer back to the appropriate unit before attempting the question again.

I am very grateful to the following Examining Groups for permission to reproduce their questions.

London and East Anglian Group (LEAG)
Midland Examining Group (MEG)
Northern Examining Association (NEA)
Northern Ireland Schools Examinations Council (NISEC)
Southern Examining Group (SEG)
Scottish Examining Board (SEB)

Unit 1

1 The diagrams below show sections of three different cells, A, B and C. They are not drawn to the same scale.

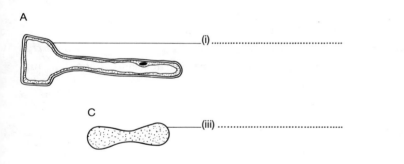

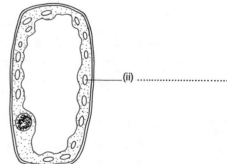

(a) Label parts (i), (ii) and (iii) on the lines provided. *(3 marks)*

(b) A and B are plant cells; C is an animal cell.
Give *two* features shown in the diagrams that support this statement. *(2 marks)*

(c) State precisely where in a plant or animal you would expect to find cells A, B and C.
(3 marks)

(d) State *one* way in which the structure of each of the cells A and B helps it to carry out its main function. *(2 marks)*

(e) (i) Name *one* substance likely to pass through cell A on its way to cell B in a living plant.
(1 mark)

(ii) Name *one* substance produced in cell B which is found in high concentration in cell C.
(1 mark)
LEAG

2 (a) Tick the correct box.
Which of the following parts of the plant cell contains chromosomes?
Cell wall ☐   Cytoplasm ☐
Nucleus ☐   Central vacuole ☐

(b) Tick the correct box.
Which of the following features is found in a plant cell but *not* in an animal cell?
Nucleus ☐   Cell membrane ☐
Cytoplasm ☐   Cell wall ☐

(c) The diagram below shows a typical animal cell.

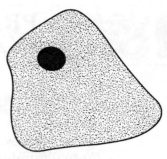

(i) Which part controls the passage of substances into and out of the cell?
(ii) Give two examples of substances which enter the cell by diffusion.

SEB

3 Three vertebrate animals are illustrated

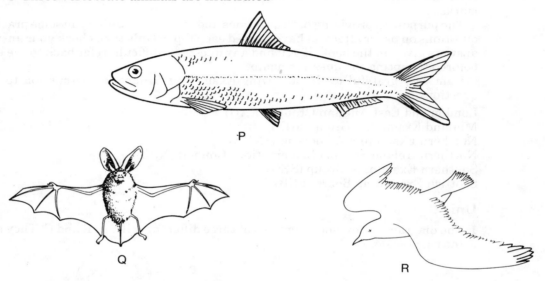

For each of the animals shown above, answer the following questions.
(a) Name the main group to which each animal belongs.   (*3 marks*)
(b) State *two* features, which can be seen in the diagram, which are characteristic of each group.
(*3 marks*)
MEG

Unit 2

4 Which of the following organisms has an exoskeleton?
A   an earthworm
B   a fish
C   a frog
D   an insect
E   a lizard

5 Which of the following is a cold-blooded animal with a dry scaly skin?
A   an amphibian
B   a bird
C   an insect
D   a mammal
E   a reptile

6 A   fungi
B   algae
C   mosses
D   ferns
E   angiosperms

Which of the plant groups listed above:
(a) lacks chlorophyll?
(b) produces seeds?
(c) are decomposers?
(d) are almost all aquatic?

NISEC

Unit 3

7 In an experiment to study the activity of bacteria, a pupil boiled some milk for 10 minutes and then left it to cool. This milk together with some raw untreated milk and sterile test tubes was used to set up the investigation as shown below.

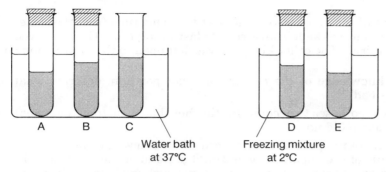

Water bath
at 37°C

Freezing mixture
at 2°C

Each sample of milk had 1 cm³ of a blue indicator dye added to it. This dye changes colour from blue to pink to colourless as the number of bacteria increase.

When examined after two days the following observations were made:

Sample	Contents	Dye colour
A	Raw, untreated milk plus dye	Colourless
B	Boiled milk plus dye	Blue
C	Boiled milk plus dye (no cover)	Pink
D	Boiled milk plus dye	Blue
E	Raw, untreated milk plus dye	Blue

(a) Which sample contained the greatest number of bacteria after two days?   (*1 mark*)

(b) Both tubes A and E contained raw, untreated milk at the start of the experiment. Suggest *one* reason why tube E showed no signs of bacterial activity.   (*1 mark*)

(c) What conclusion can be made by comparing the result obtained for tube B and tube C?   (*2 marks*)

(d) Why were tubes B and D included in the experiment?   (*1 mark*)

(e) Give *one* reason why it is important to use sterile test tubes for the milk samples.   (*2 marks*)

(f) Suggest *one* criticism which could be made of the pupil's procedure in setting up the experiment.   (*2 marks*)
    NISEC

8 An experiment was carried out to assess the sensitivity of a bacterial strain to certain antibiotics. Paper discs, each impregnated with a different antibiotic, were placed on a medium previously inoculated with the strain. The appearance of the culture, after a 24-hour period, is shown below.

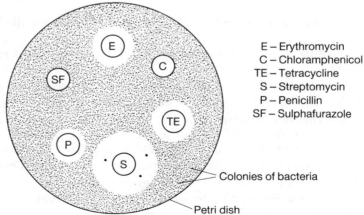

E – Erythromycin
C – Chloramphenicol
TE – Tetracycline
S – Streptomycin
P – Penicillin
SF – Sulphafurazole

Colonies of bacteria

Petri dish

(a) The effectiveness of antibiotics is normally assessed by measuring the diameter of the clear zone around each antibiotic disc. What problem might this create for the experimenter, in dealing with this set of results?

(b) What conclusions can you draw about the sensitivity of this bacterial strain to the different antibiotics?

(c) From your conclusions, what generalized statement could you make concerning the action of antibiotics on bacteria?

(d) The three colonies growing near the Streptomycin disc were thought to be resistant to this antibiotic. Describe how you would test for this hypothesis.

SEB

9 Two kilograms of tomatoes were bought at the same time from the same batch in the same shop. One kilogram was kept in an airtight plastic (polythene) bag for a week and four of these tomatoes 'went bad'. The other kilogram was left open to the air and none of these tomatoes 'went bad'.

(a) From your knowledge of the process of decomposition, describe what happens when a tomato 'goes bad'. *(5 marks)*

(b) Suggest *two* hypotheses to account for the fact that some tomatoes 'went bad' while others in the same batch did not. *(2 marks)*

(c) Choose one of your hypotheses and describe how you would carry out an experiment to test it. Give details of the apparatus you would use and of the conditions in which you would conduct your experiment. What would you measure and how would you record your results? *(8 marks)*

SEG

Unit 4

10 (a) The table shows the percentage of overweight British people in different age groups in 1981.

Age group	Percentage overweight	
	Men	Women
20–24	22	23
25–29	29	20
30–39	40	25
40–49	52	38
50–59	49	47
60–65	54	50

Source: Report of Royal College of Physicians, 1983

Use the table to answer the following questions.

    (i) Which sex in which age group has the smallest percentage overweight? *(1 mark)*

    (ii) Which age group has a greater percentage of women than men who are overweight? *(1 mark)*

    (iii) Which age group has the greatest percentage difference between men and women who are overweight? *(1 mark)*

(b) (i) What is the main substance stored in the body which forms the extra weight? *(1 mark)*

    (ii) In or around which organ is this substance likely to be found in large amounts? *(1 mark)*

    (iii) Give an example of ill health that overweight people are more likely to suffer from. *(1 mark)*

(c) (i) Explain how your body loses weight when it is at rest. *(3 marks)*

    (ii) Why do you lose more weight when you take exercise than you lose at rest? *(2 marks)*

(d) Doctors recommend a diet which includes cereals and bread with a high fibre content, vegetables and fruit, but which contains only a little fried food and only a little sugar. Give a different reason in each case why doctors recommend

    (i) a diet that includes cereals and bread with a high fibre content; *(1 mark)*

    (ii) a diet that includes vegetables and fruit; *(1 mark)*

    (iii) a diet that includes only a little fried food; *(1 mark)*

    (iv) a diet that includes only a little sugar; *(1 mark)*

SEG

11 A 22-year-old woman kept a precise record of her food and drink intake for one day. The quantities of some major nutrients were calculated, and the totals compared with the average daily requirement of a woman of that age. The figures are summarized in the table below.

Meal	Item	Quantity	Energy in kJ	Protein in g	Fat in g	Carbo-hydrate in g	Calcium in mg	Iron in mg	Vitamin C in mg
Breakfast	White bread	90 g	950	7	2	50	90	1	0
	Butter	15 g	450	0	12	0	2	0	0
	Jam	30 g	330	0	0	19	5	0	1
	Black coffee	1 cup	20	0	0	1	4	0	0
Lunch	Hamburger	150 g	1560	30	15	30	50	4	0
	Ice cream	100 g	800	4	12	20	130	0	1
	Fizzy drink	1 can	550	0	0	30	0	0	0
Evening meal	Sausages	75 g	1150	9	24	10	30	0.5	0
	Chips	200 g	2100	8	20	70	25	2	20
	Baked beans	220 g	600	10	1	20	100	2.5	4
	Apple pie	150 g	1800	5	25	60	60	1	1
	Cream	30 g	550	0.5	15	1	20	0	0
	Tea with milk	2 cups	200	2	4	6	100	0	0
Snacks	Chocolate	50 g	1200	5	20	25	120	1	0
	Peanuts	50 g	1200	15	25	5	30	1	0
TOTAL INTAKE FOR DAY			13460	95.5	175	347	766	13	27
AVERAGE DAILY REQUIREMENT			9400	58	*	*	600	14	30

*Amounts variable

(a) (i) By how much did the energy content of the day's diet exceed the average daily energy requirement of a 22-year-old woman? *(1 mark)*

(ii) What would be the probable effect of this difference on an average woman of this age over a long time? *(1 mark)*

(iii) This woman could have a daily energy requirement much greater than average. Suggest *one* reason why this would be so. *(2 marks)*

(b) The recommended carbohydrate:fat ratio in a balanced diet is 5:1 by weight.

(i) Which individual meal in the day given in the table had a carbohydrate:fat ratio of exactly 5:1? *(1 mark)*

(ii) To the nearest whole number, what is the carbohydrate:fat ratio for the whole day's intake? *(1 mark)*

(iii) What effects may this proportion of fat in the daily diet have on the woman's circulatory system? *(2 marks)*

(c) *Excluding* coffee, tea and fizzy drinks, which of the items eaten during the day had the highest level of calcium per gram of food? Give the calcium content of this item in mg per g of food. *(2 marks)*

(d) Identify *one* nutrient in which the day's food intake is *deficient* and name a deficiency disease that may result if the woman's diet continues to provide too little of the nutrient. *(1 mark)*

(e) A sample of the woman's urine taken the morning *after* the day described above contained a high concentration of urea. How might this have been predicted from the information in the table? *(1 mark)*

LEAG

Unit 5

12 A well watered geranium plant had one of its leaves covered with tinfoil as shown in the diagram below. After **three** days the leaf was removed, decolorized with ethanol and treated with iodine solution as shown.

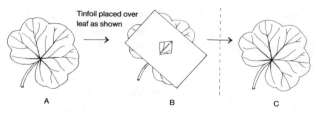

Tinfoil placed over leaf as shown

A
Geranium leaf attached to plant well watered and kept in bright light

B

C
3 days later Leaf is decolourized and treated with iodine solution.

(a) (i) What was the purpose of the tinfoil in the experiment? *(1 mark)*
(ii) Complete diagram C to show the areas stained black with iodine solution. *(2 marks)*
(iii) Name the substance in the leaf which produces the black stain with iodine. *(1 mark)*
(b) Why should the tinfoil in drawing B be fitted as shown to *both* sides of the leaf? *(1 mark)*
(c) (i) Name the coloured substance removed by the ethanol. *(1 mark)*
(ii) While the leaf is being decolorized the tube containing ethanol is not heated directly over a Bunsen burner. Explain why. *(1 mark)*
LEAG

13 The diagram below shows part of a vertical section through a green leaf.

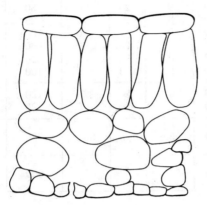

(a) On the diagram above
(i) label a guard cell and a palisade mesophyll cell; *(2 marks)*
(ii) draw a small circle *in each of the cells* that contain chloroplasts. *(3 marks)*
(b) As a result of photosynthesis, plants make sugar. State *one* use of this sugar to
(i) the plant itself; *(1 mark)*
(ii) Man *(1 mark)*
(c) The graph below shows how much sugar is produced by a green plant over a period of several days, under natural conditions.

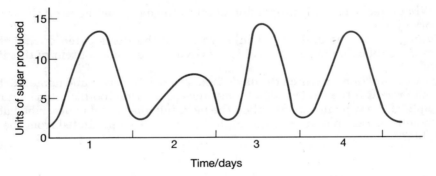

(i) Suggest *one* factor in the environment to explain why less sugar is produced during day 2. *(1 mark)*
(ii) Draw a line on the graph to show how much sugar would be produced if the plant was kept under bright light all the time. Label the line X. *(1 mark)*
MEG

14 Photosynthesis is the process by which green plants make food materials.
(a) Complete the equation which summarizes the process by writing the correct terms in the two boxes provided.

Carbon dioxide + (i) _____ $\xrightarrow[\text{sunlight}]{\text{energy from}}$ glucose + (ii) _____ *(2 marks)*

(b) What part does chlorophyll play in the process of photosynthesis? *(1 mark)*
(c) Why is photosynthesis so essential to animal life? Suggest *two* reasons. *(2 marks)*
(d) The presence of starch is often taken as an indication that photosynthesis has been taking place in leaves. The flow diagram indicates the steps taken in testing a leaf for starch.

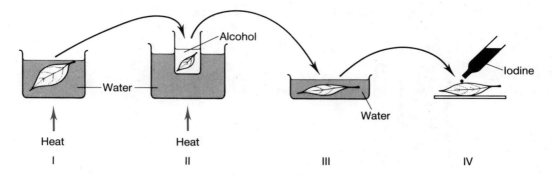

(i) What is the purpose of Stage I?

(ii) What is happening to the leaf in Stage II?

(iii) What evidence would be obtained from Stage IV which would indicate the presence of starch? *(3 marks)*

NISEC

Unit 6

15 The diagram shows a section through a human canine tooth which has begun to decay.

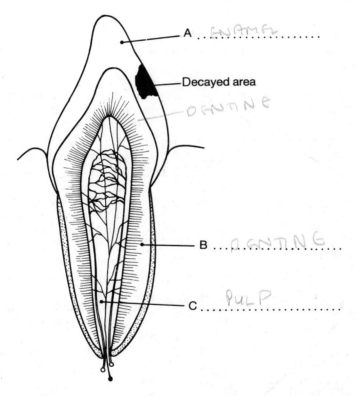

A ..ENAMEL...

Decayed area

DENTINE

B ...DENTINE...

C ...PULP...

(*a*) Name, on the diagram, the parts labelled A, B and C. *(3 marks)*

(*b*) (i) Explain how the combination of bacteria in the mouth and the eating of sugary foods may bring about dental decay. *(2 marks)*

(ii) State *two* reasons why it is important that the decay should not reach part C. *(2 marks)*

(*c*) Modern toothpastes (i) contain fluoride salts and (ii) are alkaline.

Explain how these *two* properties are likely to protect the teeth from decay. *(2 marks)*

MEG

16 (*a*) The liver and the pancreas have important roles to play in the digestion of food. Describe these roles in detail. *(10 marks)*

(*b*) The liver and the pancreas also have roles beyond that of digestion. Describe any ways in which these organs are involved in

(i) the storage of carbohydrates; *(4 marks)*

(ii) the excretion of wastes. *(6 marks)*

MEG

17 The diagram below shows a molar tooth of a horse.

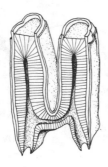

Teeth of herbivores are continually worn down by biting and chewing.

(*a*) (i) Which layer of the tooth structure is most rapidly worn down? (*1 mark*)

(ii) Label this layer clearly on the diagram above. (*1 mark*)

(iii) After a period of wear, sharp ridges appear on the surface of the tooth. What are these ridges made of? (*1 mark*)

(*b*) Suggest why the teeth of horses kept in stables may have to be filed down although the teeth of wild horses need no filing. (*2 marks*)

(*c*) Give *two* ways in which human molars differ from molars of horses. (*2 marks*)
LEAG

Unit 7

18 The graphs below show the relationship between water loss (transpiration) and water uptake in a leafy shoot.

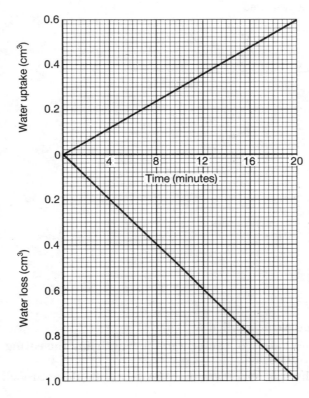

(*a*) How much water had been taken up by the plant during the first 14 minutes of the experiment? (*1 mark*)

(*b*) How much water had been lost during the same period of time? (*1 mark*)

(*c*) (i) If the shoot was kept under the experimental conditions for 12 hours, how might its final appearance differ from its appearance at the start of the experiment? (*1 mark*)

(ii) Account for the difference in appearance. (*2 marks*)

(*d*) On the axes above, sketch in the line you might expect for water uptake if the temperature around the plant was raised by 10°C, other conditions remaining constant. (*1 mark*)

(*e*) Draw a diagram of a simple piece of laboratory apparatus which you would use to investigate the rate of water uptake by a leafy shoot. (*4 marks*)
NISEC

19 (*a*) Compare, by means of concise statements, the following terms:

(i) diffusion;

(ii) osmosis;

(iii) active uptake. (*9 marks*)

(*b*) It is biologically important that certain substances enter or leave organisms through the following structures. Explain what the substances are, why they enter or leave and how the structures are adapted to the purposes they fulfil:

(i) root hairs;

(ii) leaves; (*11 marks*)

(iii) villi.
MEG

20 The graph below shows the change in mass of three detached leaves suspended under identical laboratory conditions for two hours. Oak leaves (A and B) have one surface (upper or lower) covered with vaseline. Leaf C comes from a different species of plant, and has no vaseline on it.

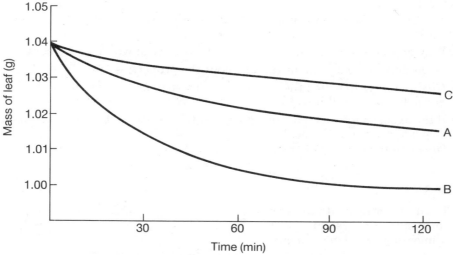

(*a*) (i) Which oak leaf has its lower surface covered with vaseline?
       (ii) Give a reason for your answer.   *(2 marks)*

(*b*) Why did leaf B stop losing mass after 90 minutes?   *(1 mark)*

(*c*) How much water would you expect an untreated oak leaf, of the same area as leaves A and B, to lose in 30 minutes?   *(1 mark)*

(*d*) In what type of habitat would you expect to find plant C? Give a reason for your answer.
                                                                              (3 marks)
                                                                              MEG

21 When a ring of all the tissues outside the xylem was removed from a woody twig (as in sketch X), it was found that the region Q above the ring became enlarged with food materials (as in sketch Y).

Which of the following is the *best* conclusion to be drawn from *this* experiment?

A a tissue outside the xylem carries food upwards

B food travels upwards in the xylem

C xylem carries food downwards

D xylem carries water and salts upwards

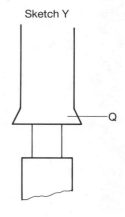

NEA

22 A piece of knotted Visking tubing was filled with a mixture of starch solution and saliva. The open end was then sealed with a tight knot and the outside of the bag washed. The bag was then placed in a beaker of water kept at 37°C. This is shown in diagram A.

A control was set up in a similar way using starch solution and boiled saliva in a Visking tubing bag. This is shown in diagram B.

After one hour the water in each beaker was tested for reducing sugar and starch.

In A the water contained reducing sugar but no starch. In B the water contained no reducing sugar and no starch.

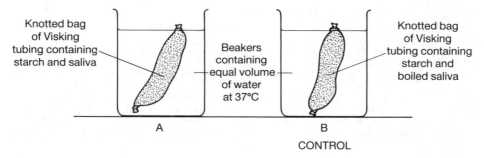

(a) Why was the bag washed before placing it in the beaker of water? *(1 mark)*

(b) Why was a temperature of 37°C chosen for this experiment? *(1 mark)*

(c) From the observations made in the experiment, what appears to be the action of saliva on starch? *(2 marks)*

(d) Describe the test you would use to detect the presence of reducing sugar in the beaker of water. *(2 marks)*

(e) Why was no reducing sugar found in the beaker of water in the control, B? *(1 mark)*

(f) What does the experiment tell us about the properties of the Visking tubing? *(2 marks)*

(g) *Give two* ways in which the Visking tubing used in this experiment is incomplete as a model of the human small intestine. *(2 marks)*

(h) If the experiment was repeated at a lower temperature, it would take longer before reducing sugar appeared in the beaker.
Give *two* reasons why this is so. *(2 marks)*
LEAG

Unit 8

23 The diagram opposite shows part of the respiratory system and alimentary canal in a human.

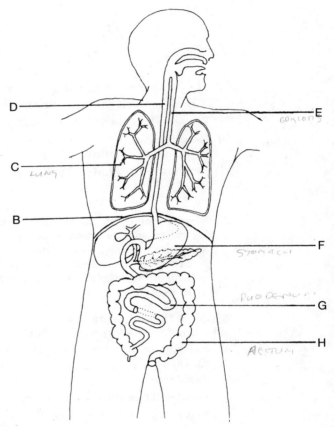

(a) Different substances pass into the blood at different places. Name *one* substance passing into the blood at each of C, G and H. *(3 marks)*

(b) Name *one* substance excreted through E. *(1 mark)*

(c) Give *one* result in each case of muscle contraction at B, D and F. *(3 marks)*

(d) Certain substances may enter the body and have harmful effects.
Name *two* harmful substances that may enter the body through D and *two* others that may enter through E.
Describe *one* harmful effect of each substance named. *(4 marks)*

Entry through		Harmful substances	Harmful effects
D	(i)		
	(ii)		
E	(i)		
	(ii)		

(Unit 6 may also be helpful if guidance is needed whilst answering this question)
LEAG

24 Most of the carbon dioxide in the blood is carried:

A   as carboxyhaemoglobin in red blood cells
B   by the white blood cells
C   as bubbles of gas in the plasma

D   in association with the platelets
E   as hydrogencarbonate (bicarbonate) in the plasma.

NISEC

25 The diagram opposite shows a mammalian heart.

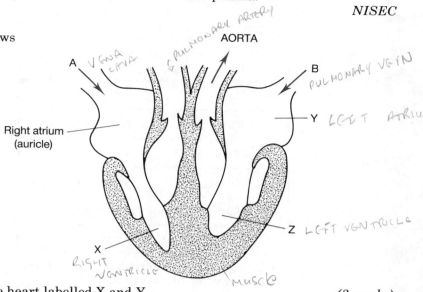

Right atrium (auricle)

A — VENA CAVA
PULMONARY ARTERY
AORTA
B — PULMONARY VEIN
Y — LEFT ATRIUM
Z — LEFT VENTRICLE
X — RIGHT VENTRICLE
MUSCLE

(*a*) Name the chambers in the heart labelled X and Y. (*2 marks*)
(*b*) Name the *two* blood vessels labelled A and B which bring blood to the heart. (*2 marks*)
(*c*) State *two* differences in the composition of the blood in blood vessels A and B. (*2 marks*)
(*d*) Why is the wall of the heart thicker in chamber Z than in chamber X? (*2 marks*)
(*e*)   (i) Through which blood vessels does the heart muscle receive its own blood supply? (*1 mark*)
CORONARY ARTERIES
   (ii) Why should a blockage in these vessels cause a 'heart attack'? (*4 marks*)
The oxygen supply to the muscle (cardiac) is stopped and the muscle ceases to work well
NISEC

26 The diagram below, labelled A to K, shows a ventral view of a mammalian heart and blood vessels.

Right Side   Left Side

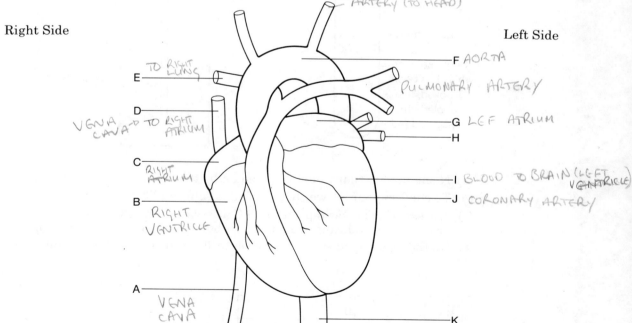

ARTERY (TO HEAD)
E — TO RIGHT LUNG
D — TO RIGHT ATRIUM
VENA CAVA
C — RIGHT ATRIUM
B — RIGHT VENTRICLE
A — VENA CAVA
F AORTA
PULMONARY ARTERY
G — LEFT ATRIUM
H
I BLOOD TO BRAIN (LEFT VENTRICLE)
J CORONARY ARTERY
K
AORTA

(*a*) State *precisely* where the blood in D and E would go to next. (*2 marks*)
(*b*) Write the letter of the part which contracts to send blood to the brain. (*1 mark*)
(*c*)   (i) What effect may heavy smoking have on J? (*1 mark*)
   (ii) How may this affect the heart as a whole? (*1 mark*)
LEAG

Unit 9

27 The table below shows the winning times in races held at three Olympic Games at three different cities.

Race	Tokyo – 1964 200 m above sea level	Mexico – 1968 2240 m above sea level	Munich – 1972 52 m above sea level
100 m	10.0 s	9.9 s	10.2 s
200 m	20.3 s	19.8 s	20.0 s
400 m	45.1 s	43.8 s	44.7 s
800 m	1 min 45.1 s	1 min 44.3 s	1 min 45.9 s
1500 m	3 min 38.1 s	3 min 34.9 s	3 min 36.3 s
5000 m	13 min 48.8 s	14 min 28.4 s	13 min 26.4 s
10 000 m	28 min 24.4 s	29 min 27.4 s	27 min 38.4 s

(*a*) Using only the information in the table, what are the effects of height above sea level on performance of the athletes? *(2 marks)*

(*b*) State *two* pieces of evidence which support your answers to section (a). *(2 marks)*
NEA

28 To investigate anaerobic respiration in yeast, the apparatus shown in the diagram below was set up:

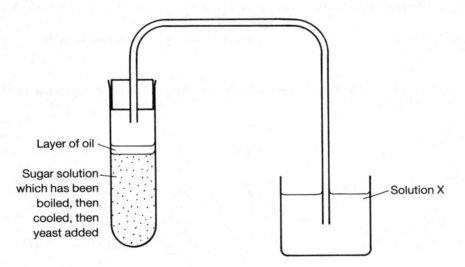

Layer of oil

Sugar solution which has been boiled, then cooled, then yeast added

Solution X

(*a*) Name the gas that would be given off by the yeast. *(1 mark)*

(*b*) To confirm your answer to (a), what is solution X? *(1 mark)*

(*c*) The experiment was repeated at 5°C intervals from 25°C to 65°C. The rate of respiration was measured by counting the number of bubbles of gas given off per minute.
On the axis provided, draw a graph to show the results you would expect: *(2 marks)*

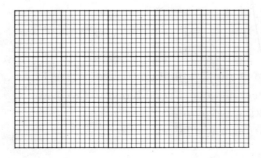

(*d*) Explain why your graph has this particular shape.   *(2 marks)*

(*e*)   (i)   How would the amount of ATP produced by the respiration of the yeast compare under aerobic and anaerobic conditions?   *(1 mark)*

(ii)   How will this difference in ATP production affect the growth of the yeast?   *(3 marks)*

(*f*)   (i)   Why are a yeast, and a mould such as *Penicillium*, both classified as fungi?   *(2 marks)*

(ii)   What is the principal difference between a yeast and a species of *Penicillium*?   *(1 mark)*

(Unit 2 may also be helpful if guidance is needed whilst answering this question)   *MEG*

29 When a peanut is burnt under a test tube of water, as shown in the diagram, the water heats up.

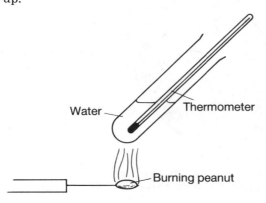

(*a*) What is being measured in this experiment?   *(1 mark)*

(*b*) The temperature of the water needs to be measured before and after the peanut is burnt under the test tube. Why is this necessary?   *(2 marks)*

(*c*) State *two* ways in which the result of the experiment may be inaccurate.   *(2 marks)*
MEG

30 (*a*) Use the following list of words in a sentence, or sentences, to explain the term *aerobic respiration*.

sugar, oxygen, energy, carbon dioxide   *(2 marks)*

(*b*) A muscle is working as hard as possible.

(i)   What change would occur to its type of respiration, compared to that in a resting muscle?   *(1 mark)*

(ii)   What new product is made?   *(1 mark)*
MEG

Unit 10

31 The graph shows the mean body temperatures of two groups of men over a period of four days. Both groups did the same work. One group lived in a cool climate, the other in a hot climate.

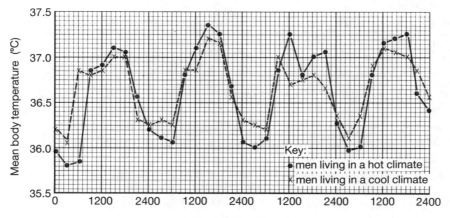

Write down *five* things that the graph tells you about changes in body temperature.   *(5 marks)*
SEG

Source: *Man — Hot and Cold*, by Otto G. Edholm

32 The diagram shows a section through a mammal's kidney.

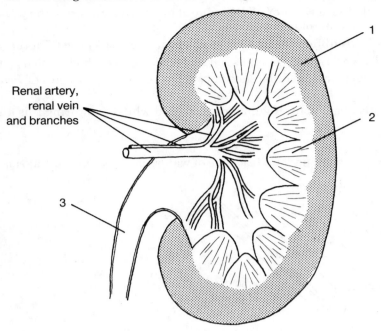

(a) Name the parts 1, 2 and 3. *(3 marks)*

(b) The table gives information about the human kidney.

Rate of blood flow in kidneys	Rate of filtration into kidney tubules (nephrons)	Rate of urine passing out of kidneys
1.2 dm³ per minute	0.12 dm³ per minute	1.5 dm³ per day

(i) What percentage of blood passing into the kidney is filtered into the kidney tubules? *(1 mark)*

(ii) Where in the kidney (part 1, 2 or 3 on the diagram) does filtration take place? *(1 mark)*

(iii) About 172 dm³ are filtered from the blood into the kidney tubules per day, yet only 1.5 dm³ of urine are excreted. What happens to the other 170.5 dm³? *(2 marks)*

(c) The table shows the average amounts of urine, sweat and salt (sodium chloride) lost on a normal day, a cold day and a hot day. (Assume that food and drink are the same on all days.)

	Urine lost per day (dm³)	Sweat lost per day (dm³)	Salt (sodium chloride) lost per day	
			in urine (g)	in sweat (g)
Normal day	1.5	0.5	18.0	1.5
Cold day	2.0	0.0	19.5	0.0
Hot day	0.375	2.0	13.5	6.0

(i) Why is more urine lost on a cold day than on a normal day? *(2 marks)*

(ii) Why do you think the total amount of salt lost on each of the three days is the same? *(2 marks)*

(iii) The minimum amount of urine excreted in a day is 0.375 dm³. Why do you think the kidneys always produce some urine? *(2 marks)*

(iv) What *must* someone losing more than 7 dm³ of sweat in a day do in order to remain healthy? *(2 marks)*

SEG

33 Explain how urea passes from the liver, where it is made, to the bladder. *(14 marks)*

LEAG

34 The diagram below shows a vertical section of human skin.

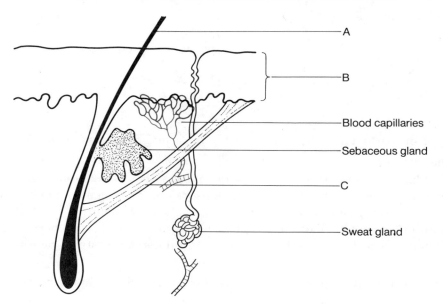

(a) Name the parts A, B and C. *(3 marks)*

(b) After running a race, your skin is wet and your face is hot.
    (i) Why is your skin wet and how does this help to cool your body? *(4 marks)*
    (ii) Why is your face hot and how does this help to cool your body? *(4 marks)*

(c) How do you think the sebaceous glands help to keep hair and skin healthy? *(3 marks)*
                                                      LEAG

Unit 11

35 The diagram represents a section through the human eye.

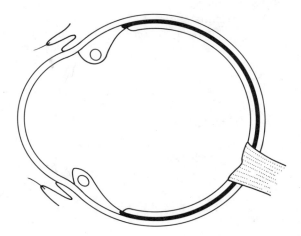

(a) Complete the diagram by drawing in the correct position of the lens, the suspensory ligament, the iris and the pupil. *Label each clearly.* *(4 marks)*

(b) Use the letter B to label the area of the retina which does not contain light sensitive cells. *(1 mark)*

(c) State *two* ways in which a person's sight would be affected by the loss of one eye. *(2 marks)*

(d) A football spectator was seated in the grandstand reading the match programme when the players ran on to the pitch. What changes would have to occur in the focusing mechanism of the spectator's eyes to enable the players to be seen clearly? *(3 marks)*

(e) Some evidence indicates that many sunglasses were not doing their job of cutting out the harmful ultra-violet rays in the sunlight. The lenses may look dark but wearing these glasses may be even more damaging than wearing none at all.
    (i) How would the appearance of the pupil behind the dark lenses differ from its appearance in bright light? *(1 mark)*
    (ii) How do the muscles in the iris bring about the change noted above? *(2 marks)*
    (iii) Why could wearing these defective glasses be more damaging than not wearing any at all? *(2 marks)*
                                                      NISEC

Unit 12

36 The diagram below shows a motor neurone.

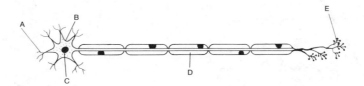

Which of the parts labelled, A, B, C, D or E, transmits impulses to a muscle or gland?

37 What is the sequence of events in a reflex action?

A   stimulus – effector – receptor – sensory neurone – central nervous system – motor neurone
B   stimulus – effector – motor neurone – sensory neurone – central nervous system
C   stimulus – receptor – central nervous system – sensory neurone – motor neurone – effector
D   stimulus – receptor – motor neurone – sensory neurone – central nervous system – effector
E   stimulus – receptor – sensory neurone – central nervous system – motor neurone – effector

MEG

38 In an experiment three plant shoots were treated as shown in the diagram below.

Lateral bud

Lanolin paste

Lanolin paste + IAA (a plant hormone)

SHOOT A
Normal shoot control

SHOOT B
Tip of stem cut off and coated with lanolin paste on cut surface

SHOOT C
Tip of stem cut off lanolin paste + IAA placed on cut surface

After a few days the appearance of the shoots was as shown in the diagram below.

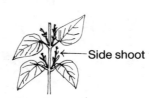

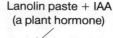

Side shoot

SHOOT A   SHOOT B   SHOOT C

(a) What appears to be the effect of removing the growing tip from the shoot B?   *(1 mark)*
(b) Explain why plain lanolin paste should be placed on shoot B?   *(1 mark)*
(c) What can you deduce about the function of the growing tip as a result of this experiment?   *(1 mark)*

LEAG

39 The diagram below represents a simple reflex arc.

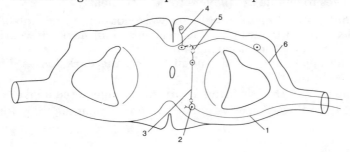

(a) Which one of the following labels the sensory neurone?

   A 1
   B 2
   C 4
   D 5
   E 6

(b) Which one of the following shows the path followed by an impulse in a reflex arc?

Direction of impulse

   A $6 \rightarrow 4 \rightarrow 5 \rightarrow 3 \rightarrow 1$
   B $6 \rightarrow 5 \rightarrow 3 \rightarrow 2 \rightarrow 1$
   C $1 \rightarrow 3 \rightarrow 4 \rightarrow 5 \rightarrow 6$
   D $1 \rightarrow 2 \rightarrow 3 \rightarrow 5 \rightarrow 4$
   E $1 \rightarrow 2 \rightarrow 3 \rightarrow 5 \rightarrow 6$

Unit 13

40 The photograph is of a hind-leg bone of a sheep cut in half lengthwise. Look at the photograph carefully and describe *three* ways in which the structure of this bone helps in either support or movement of the sheep's body. *(6 marks)*

                                                      SEG

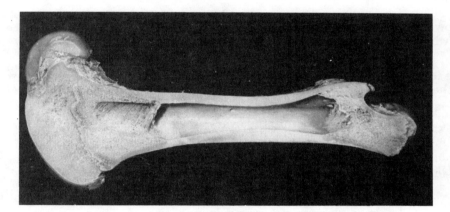

41 The diagram below shows the main parts of the human arm used to lift an object.

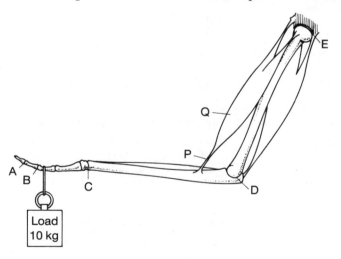

(a) What is the name of the tissue labelled P?

   A muscle   D tendon
   B ligament   E nerve
   C cartilage

(b) Which one of the joints labelled A,B,C,D or E is a slipping joint?

(c) The hand can just support the load shown in the diagram. What is the tension in the muscle Q?

   A much less than 10 kg   D approximately 2 kg
   B much greater than 10 kg E less than 2 kg
   C approximately 10 kg

Unit 14

42

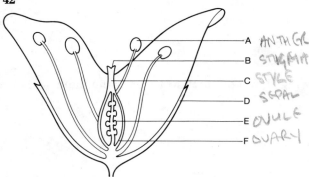

A ANTHER
B STIGMA
C STYLE
D SEPAL
E OVULE
F OVARY

(a) Using the diagram of the half-flower, name the structures A–F.   *(6 marks)*

(b) What is the function of the structure labelled B?   *(1 mark)*

(c) What does structure F give rise to after fertilization?   *(1 mark)*

(d) State *two* features which suggest that the flower is insect-pollinated.   *(4 marks)*

(e) Explain how a gardener could prevent self-pollination in a flower of the type shown above.
(2 marks)

(f) Flowers of wind-pollinated trees bloom before the leaves appear on the tree. State *one* way in which this is an advantage to the plant.   *(2 marks)*

NISEC

43 The following events occur during sexual reproduction in a flowering plant. Arrange them in the correct order by writing the appropriate letter in the boxes in the table below.

A   meiosis takes place in the anther

B   pollen tube grows down into the ovary

C   a male nucleus fuses with the ovum nucleus

D   anthers split open releasing pollen

E   insects transfer pollen from anther to stigma

Order of events	Statement letter
1	
2	
3	
4	
5	

44 The diagram below shows the external appearance of a mung-bean seed.   *MEG*

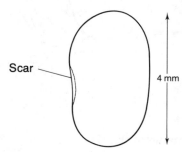

Scar

4 mm

(a) From what part of the flower has the seed developed?   *(1 mark)*

(b) You are asked to find the best temperature for sprouting (germinating) mung-bean seeds. You are given 100 seeds, 10 test tubes, cotton wool, 10 thermometers, a clock and a supply of water.

  (i) In the space below draw and label one of the test tubes you would set up for the experiment.   *(2 marks)*

  (ii) Explain how you would carry out this experiment.   *(7 marks)*

  (iii) Describe *one* difficulty you might have in carrying out this experiment.   *(1 mark)*

LEAG

45 In late summer young stems of willow were cut into 30 cm lengths. The lower 10 cm of each stem cutting were stripped of leaves and buds and then buried firmly out of doors in coarse gritty soil, as shown in the diagram.

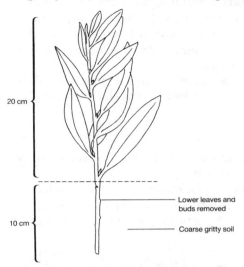

After a few weeks all the leaves dropped off. After a few months roots formed on the cut stem ends. The following spring 70% of the stem cuttings grew into new plants.

(a) (i) Suggest one reason for removing the leaves and buds from the lower 10 cm of stem.
(1 mark)

(ii) Suggest one reason for using a coarse gritty soil instead of a fine soil.   *(1 mark)*

(iii) Suggest one reason for burying each stem firmly (by pressing down the soil around it).
(1 mark)

(b) (i) Suggest one way in which loss of all the leaves after a few weeks might have helped the rooting of the stems.   *(2 marks)*

(ii) Suggest one way in which loss of all the leaves after a few weeks might have hindered the rooting of the stems.   *(2 marks)*

(c) (i) Describe an improvement on this method of growing stem cuttings.   *(1 mark)*

(ii) Explain how your improvement would work.   *(1 mark)*

(d) A year after the cuttings were put in the soil one of the new plants was 1.2 m tall with six side branches and another was 0.58 m tall with eleven side branches.

(i) Describe an experiment you could carry out during the following year to help decide whether the differences between these two plants were caused by inheritance or had some other cause.   *(4 marks)*

(ii) What results of your experiment would suggest that the differences were inherited?
(1 mark)

(iii) What results of your experiment would suggest that the differences had some other cause?   *(1 mark)*
SEG

Unit 15

46 The diagram below shows a foetus in the womb.

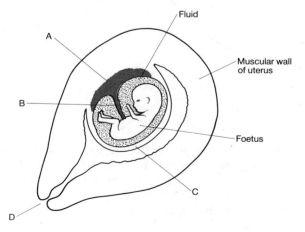

(a) Name the parts labelled A, B, C and D.   *(4 marks)*

(*b*) The growing foetus needs a supply of oxygen. Describe clearly how it gets this. *(2 marks)*

(*c*) What is the function of the fluid which surrounds the foetus? *(1 mark)*

(*d*) Describe how the muscles in the uterus wall help the birth of the baby. *(2 marks)*

(*e*) Concern has been expressed at the number of children suffering from Down's syndrome being born to families living along the east coast of Northern Ireland.

    (i) What may have happened in the mother's body which could have been responsible for the condition appearing in the child? *(1 mark)*

    (ii) How would the chromosomes in a cell taken from a Down's syndrome baby differ from those taken from a baby who does not suffer from the condition? *(1 mark)*

                  NISEC

47 (*a*) (i) Oestrogen is a mammalian sex hormone. Name the organ which secretes it and explain why the secretion is called endocrine. *(3 marks)*

    (ii) Name the mammalian male sex hormone and the organ which secretes it. *(2 marks)*

(*b*) The ripening of an egg cell prior to ovulation is associated with a high output of oestrogen. Prolactin is a hormone secreted by the pituitary gland in relatively high concentration during lactation (breast feeding).

    The following graphs for lactating women show mean levels of suckling frequency and blood hormonal concentrations for the same time intervals after birth.

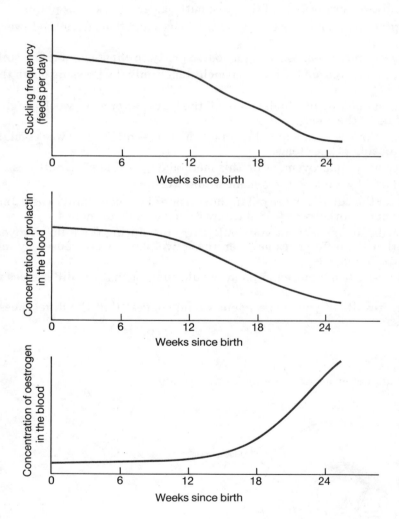

Explain carefully how the above information helps to explain the following:

    (i) Among some traditional hunter-gatherer people in the Third World it is the custom to breast feed infants for several years and it is common for the interval between successive births in a family to be four or more years. *(6 marks)*

    (ii) Infertility clinics sometimes find that their women patients have abnormally high blood concentrations of prolactin. *(4 marks))*

                  NISEC

Unit 16

48 The curves in the graph show the growth of two tomato seedlings, A and B, over a period of 20 days.

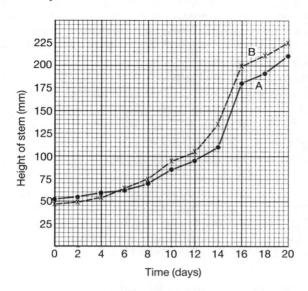

The tomato seedlings were grown in pots filled with soil. The pots were placed on the same window-sill in a warm room during the experiment. The soil of seedling A was treated with 300 cm³ distilled water at the beginning of the experiment and every five days after that. The soil of seedling B was treated with 300 cm³ of a suitably diluted liquid fertilizer at the beginning of the experiment and every five days after that.

(*a*) What was the height of each stem at the beginning of the experiment?
  Plant A .... mm
  Plant B .... mm   *(2 marks)*

(*b*) What was the height of each of the stems on day 17 of the experiment?
  Plant A .... mm
  Plant B .... mm   *(2 marks)*

(*c*) During which two-day period was the growth of seedling A most rapid?   *(1 mark)*

(*d*) Calculate the growth rate per day for seedling A over this period. (Show your working.)
                                                                          (2 marks)

(*e*) It is not certain that the faster growth of seedling B is due to the liquid fertilizer. Suggest three possible explanations for the faster growth.   *(3 marks)*
                                                                              MEG

49 The diagram below shows the results of an experiment to determine the effects of various treatments on the germination of mustard seeds, and their subsequent development over a period of nine days.

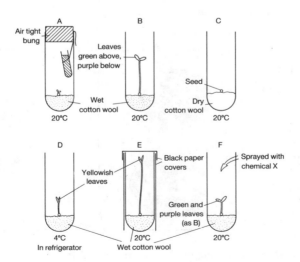

Ten seeds were dropped onto the cotton wool in each tube, A to F. All the tubes except C were placed in a rack, on a bench by a window, in full light.

The one seedling shown in each tube is drawn to scale, and represents the average height of all the seedlings that grew.

Two other tubes, G and H, were also set up but these are not shown in the diagram. They were set up in the same way as tubes B and C and were put in the same rack. However, both tubes G and H contained soil from a compost heap instead of cotton wool. The soil in G was wet but in H it had first been oven-dried. When inspected after 9 days, the seedlings in G had withered and died whereas the seeds in H had not even germinated. Soil G was still damp.

Now answer the following questions.

(*a*) From your observations of tubes A to E only, what conclusions can be made regarding the conditions necessary for the germination of these mustard seeds?   *(2 marks)*

(*b*) What do you conclude from the results in tube A about the nature of the liquid in the small test-tube? Explain why growth of this seedling is so poor.   *(4 marks)*

(*c*) Explain fully, with reference to biological principles,
    (i) the appearance of the seedlings in tubes B and E;
    (ii) seedling sizes in tubes D and E.   *(8 marks)*

(*d*) What can you conclude from the results obtained by the use of tubes G and H in this experiment? Give an explanation of the results.   *(3 marks)*

(*e*) When the biomass of the seedlings from B and from E were compared they were found to be very similar. However, when both sets of seedlings had been oven-dried, the biomass of seedlings in B was clearly greater than that from E.
Explain these results.   *(4 marks)*

(*f*) Chemicals such as X are used as sprays to increase the yield of wheat in Britain. They do not discourage pests such as herbivores nor diseases such as rusts, nor do they affect soil conditions. They are particularly useful in East Anglia, where many hedgerows have been removed.

Suggest how the use of chemicals such as X might increase wheat yields.   *(2 marks)*
                                                                                        MEG

Unit 17

50 The following diagrams show stages in mitosis. Using the identifying letters arrange these stages in the correct sequence.

U

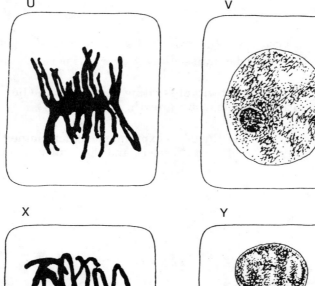

V

W

X

Y

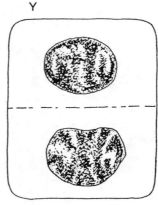

Z

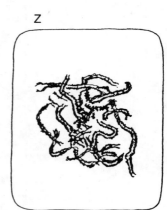

51 DNA is a molecule containing a number of alleles arranged lengthwise.
(*a*) Where would you find DNA inside a cell?   *(1 mark)*
(*b*) Why is DNA such an important substance?   *(2 marks)*

(c)   (i) What is an allele? *(1 mark)*
      (ii) Give one example of an allele found in humans. *(1 mark)*
      SEG

52 The drawing below shows the results of a cross between two fruit flies.

There are four types of fruit flies in the above drawing.

Male Long   Male Short   Female Long   Female Short

You have to count the numbers of each type of fruit fly produced by the cross.

(a) Explain how you counted the four types of fruit flies. *(1 mark)*
(b) Make a suitable table to show the numbers of each type of fruit fly. *(4 marks)*
(c) What were the genotypes of the parent fruit flies? *(1 mark)*
NEA

53 Examine the diagram below and answer the following questions:

(*a*) Given that the number of chromosomes in human body cells is 46, enter in the circles on the diagram the number of chromosomes for each of the structures shown.   *(4 marks)*

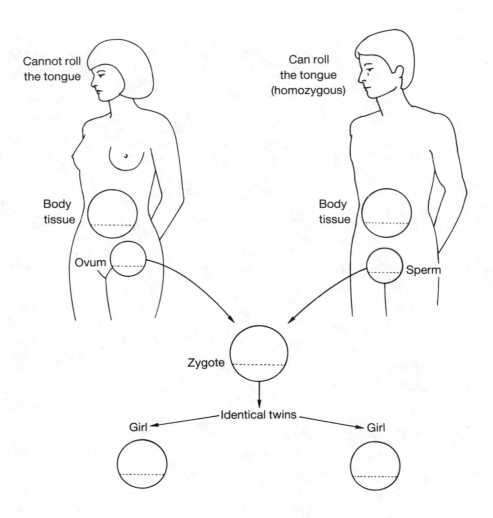

(*b*) A scientist carried out four crosses of true-breeding pea plants. Each cross showed contrasting characters.

The F_1 plants which resulted were then allowed to self-pollinate.

The table summarizes the four crosses. It shows each of the original parental crosses and the resulting F_1 plants. It also shows the F_2 plants which resulted from the self-pollination of the F_1 generation and the F_2 ratios.

Cross number	Original parental cross	F_1 plants from parental cross	F_2 plants from F_1 cross	F_2 ratio
1	Tall × short	All tall	787 tall : 277 short	2.84 : 1
2	Round seeds × wrinkled seeds	All produce round seeds	5474 round : 1850 wrinkled	2.96 : 1
3	Yellow cotyledons × green cotyledons	All produce yellow cotyledons	6022 yellow : 2001 green	3.01 : 1
4	Grey seed coat × white seed coat	All produce seeds with grey coats	705 grey : 244 white	

(i) Complete the following table to show the dominant and the recessive characters in each of the crosses.   *(2 marks)*

Cross number	Dominant character	Recessive character
1		
2		
3		
4		

(ii) Explain how you can tell from the table of results which character is dominant and which is recessive. *(2 marks)*

(c) (i) Calculate the F_2 ratio of plants producing grey-coloured seed to plants producing white-coloured seed in cross 4. *(Show your working.)* *(2 marks)*

(ii) To the nearest whole number, what is the ratio of dominant to recessive characters in all four crosses? *(1 mark)*

(d) (i) Using the letters T and t to represent genes for tallness and shortness, enter in the circles the genotypes of each parent in cross 1. *(1 mark)*

Tall parent   ×   Short parent

(ii) All the offspring (F_1 plants) resulting from this cross are tall plants. State the genotype of the F_1 plants. *(1 mark)*

(iii) Complete the diagram to explain the F_2 results obtained in cross 1. *(3 marks)*
MEG

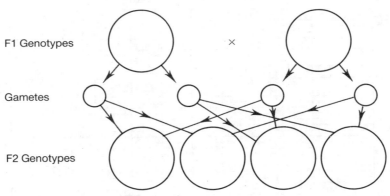

F1 Genotypes   ×

Gametes

F2 Genotypes

54 The table shows the approximate risk that women of different ages will have a baby with Down's syndrome.

Mother's age (years)	Approximate risk per 10 000 births
17	4
22	6
27	8
32	11
37	34
42	100
47	217

Source: Dr J.L. Hamerton

(a) Draw a graph of these figures using the graph paper and the scales on p.194. Join the points that you plot with straight lines. *(4 marks)*

(b) What is the risk that a baby born to a woman of 40 will have Down's syndrome? *(1 mark)*

(c) How many times greater is the risk that a woman will have a Down's syndrome baby if she is 42 than if she is 22? *(1 mark)*

(d) Suggest two reasons why babies with Down's syndrome are more often born to older mothers. *(2 marks)*
SEG

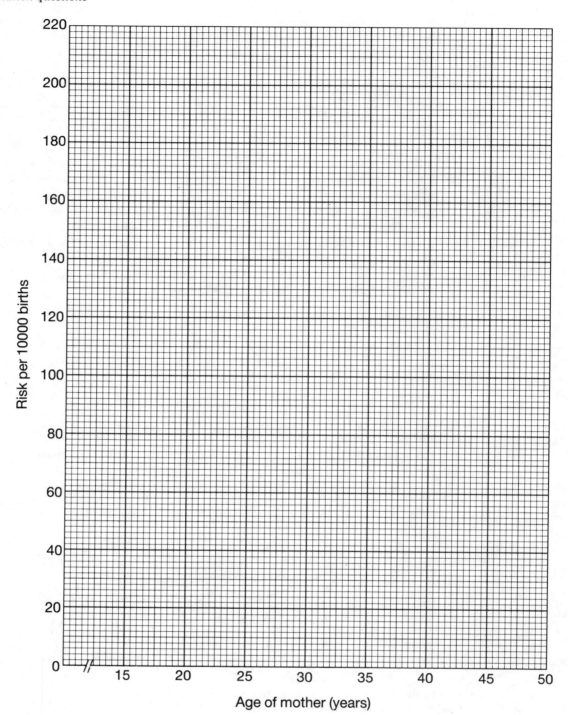

Risk per 10000 births

Age of mother (years)

55 (*a*) Describe briefly what is abnormal about the blood of a haemophiliac.   *(2 marks)*

(*b*) X^hY is a common way of representing the genotype of a haemophiliac. List three important pieces of information that these symbols convey.   *(6 marks)*

(*c*) It is very unlikely that a person of genotype X^hY would have haemophiliac children, yet there is a considerable chance of a grandchild being haemophiliac. Using simple genetical diagrams, justify the above statement.   *(6 marks)*
NISEC

Unit 18

56 A moth trap was set up on a school roof in the middle of a large industrial town. In an experiment using this moth trap, a group of pupils collected several peppered moths. Most of the moths were similar to the ones labelled A but a few resembled moth B.

(*a*) Give two distinct differences between the appearance of the wings of the moths labelled A and B.   *(2 marks)*

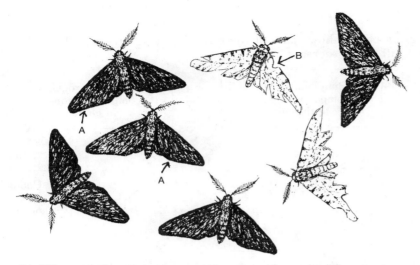

(b) When moths rest, they settle on surfaces such as trees or walls with their wings out-stretched.

The students decided to see whether moths like type A survived better than moths looking like moth B. Small marks were made on the undersides of the trapped moths' wings, then the moths were released. A few days later the moth trap was set up again and the catch was examined to see how many marked moths were recaptured.

The results are shown below.

	Number of moths marked and released	Number of marked moths which were recaptured
Type A	416	119
Type B	168	22

(i) Why were the moths marked on the undersides of their wings? *(1 mark)*
(ii) Which type of moth survived better in the industrial town? *(1 mark)*
(iii) Explain why this type of moth survived better. *(2 marks)*
LEAG

57 The kiwi fruit originally grew wild in China. In the 1960s a New Zealand farmer planted a number of seeds intending to select plants with the best features.

(a) What name is given to this form of selection? *(1 mark)*

(b) Why was it important for the farmer to use seeds from a large number of different plants? *(1 mark)*

(c) The best variety is called Hayward after the farmer who discovered it. Most of the world's cultivated kiwi fruit is now the Hayward variety.

(i) How must Hayward plants be reproduced to keep the variety the same? *(1 mark)*
(ii) What could research workers do if they want to improve on the Hayward variety? *(2 marks)*
LEAG

58 In parts of the world where malaria is common, people suffering from sickle-cell anaemia are also found. Blood from people from a malarial area and a non-malarial area was tested both for type of red blood cells and for the presence of malarial parasites: the results are given below.

Person's blood	Malarial parasites present
Normal cells	47%
Sickle cells	29%

(a) What effect does the person's type of blood cells have on their chances of having malarial parasites present? *(1 mark)*

(b) Would you expect people with normal or with sickle-cell blood (but not anaemic) to survive longer in a malarial area? Give a reason for your answer. *(2 marks)*
MEG

Unit 19

59 The diagram below shows part of the nitrogen cycle in a woodland. The labelled arrows represent different processes.

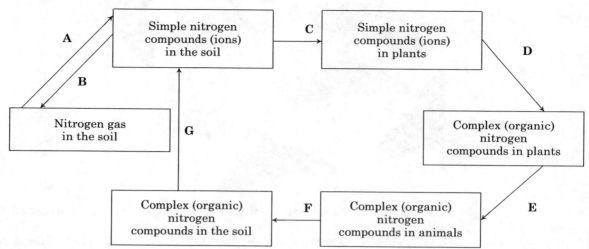

In the table below, write the name of a process occurring at each labelled arrow. Two lines have been completed for you.

Arrow label	Process
A	
B	denitrification
C	
D	protein synthesis
E	
F	
G	

(5 marks)
LEAG

60 The food web was constructed following a study of a grassland area.

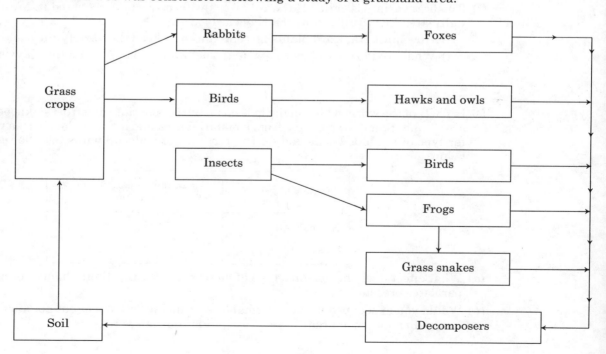

(*a*) From the food web select
   (i) a herbivore
   (ii) a carnivore
   (iii) a secondary consumer   (*3 marks*)

(*b*) What role is played by the decomposers in maintaining the food web?   (*2 marks*)

(*c*) After the area had been sprayed with a pesticide to control insect larvae several small birds were found to have died from chemical poisoning. Suggest how this may have occurred.
          (*2 marks*)

(*d*) Some farmers have removed hedgerows and small wooded areas to increase the amount of land used to produce crops. The crops in these larger fields can then be harvested with large modern machinery. How would this action affect the hawks and owls in the area being studied?   (*3 marks*)
          NISEC

61 The following table relates to the feeding and growth of stick-insects through a series of growth or nymphal stages. All the figures are averages for six insects and each lasted approximately the same time.

		Growth stage (or nymphal stage)						
		1	2	3	4	5	6	
Cumulative dry mass (mg) by the end of each stage with respect to:	Privet leaves consumed	30	2	80	105	160	380	500
	Food assimilated	22	34	56	73	105	224	258
	Food respired	17	26.5	46	58	83	179	213
	Faeces produced	8	16	24	32	55	156	242
	The stick-insect	5	7.5	10	15	22	45	45

(*a*) (i) Like many other insects, stick-insects moult at the end of each nymphal stage. Describe briefly what happens when an insect moults and explain why moulting is an essential feature of its growth.   (*3 marks*)
   (ii) State *two* other characteristics of insects as a group.   (*2 marks*)

(*b*) Values for food assimilated and food respired have been calculated from the other data provided. Briefly explain the method of calculation for both of these using the figures for the first growth stage.   (*4 marks*)

(*c*) Compare average insect dry masses at the end of growth stages 1, 4 and 6. What do they indicate about the rate of growth in the later nymphal stages compared with the earlier? Assume each growth period takes the same time.   (*4 marks*)

(*d*) (i) Name the trophic levels occupied by privet and stick-insect.   (*2 marks*)
   (ii) For the adult stage calculate the value of

$$\frac{\text{Dry mass of stick-insect}}{\text{Total dry mass of leaves consumed}} \times \frac{100}{1}\%$$ (*2 marks*)

   (iii) Assume that a similar value would be arrived at for animals reared for food and use it to explain the biological reasoning behind the following quotation: 'To help the world's food supply people, especially in the developed world, should eat more plant and less animal produce'.   (*4 marks*)
          NISEC

62 The diagram below shows a section through a pond where two samples of animals, sample A and sample B, were collected.

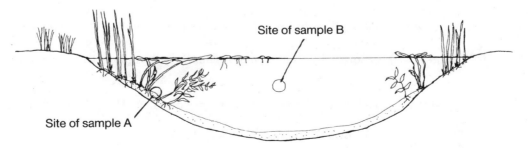

Site of sample B

Site of sample A

The table shows the animals collected in each sample. The same method was used to take samples at each site.

Animal	*Number of animals	
	Sample A	Sample B
Snails	90	2
Mites	140	80
Leeches	5	2
True worms	80	0
Flatworms	10	1
Insects — Damsel-fly nymphs	30	5
Water boatmen	170	45
Mayfly nymphs	50	100
Midge larvae	120	35
Beetles	30	15

* (Numbers simplified from actual data)

(a) (i) Which animal was present in the largest number at site A?   (*1 mark*)

   (ii) Which animal was present in the largest number in the combined samples, A and B?
                                                                                   (*1 mark*)

(b) Complete the circle below to form a pie-chart of the insects at site A. The circle has been divided into 20 equal parts. The sector for the damsel-fly nymphs has been completed on the pie-chart to help you.   (*3 marks*)

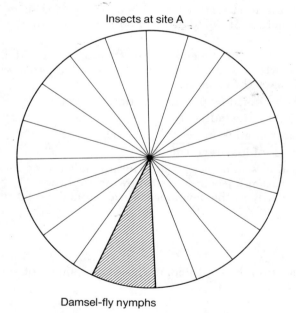

Insects at site A

Damsel-fly nymphs

(c) From the numbers given in the table, which animal is likely to be a secondary consumer?
                                                                                   (*1 mark*)

(d) Suggest *two* reasons why there are more snails in sample A than in sample B.   (*2 marks*)
                                                                                   LEAG

63 The drawings below show five different flatworms, A, B, C, D and E which are found in freshwater.
   The key which follows can be used to identify them.

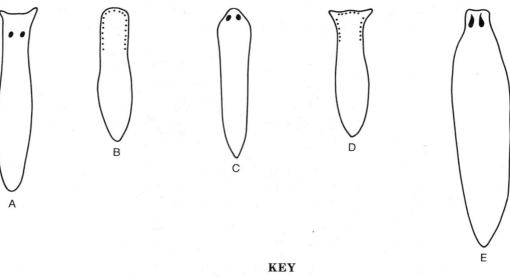

KEY

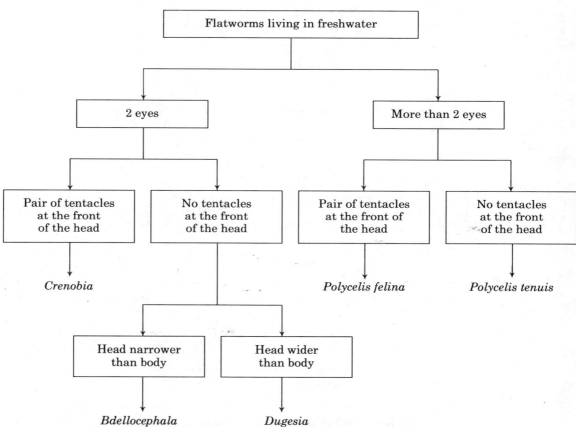

Use the key to identify the flatworms shown in the drawings A to E.
Write the letter of each flatworm next to its name below.
Bdellocephala
Crenobia
Dugesia
Polycelis felina
Polycelis tenuis

(*5 marks*)
LEAG

64 (*a*) Describe *one* method you could use to investigate the distribution of a named animal in a deciduous wood. (*5 marks*)

(*b*) Explain the influence that humans and other animals living in the area surrounding a wood might have on the woodland ecosystem. (*5 marks*)

(*c*) Explain how acid rain is caused and what effects it might have on woodland. Suggest ways in which acid rain could be reduced. (*4 marks*)
LEAG

65 The diagram below shows a section through a wood where two samples of animals, sample A and sample B, were collected.

Site of sample B

Site of sample A

The table below shows the animals collected in each sample. The same method was used to take samples at each site.

Animal	*Number of animals	
	Sample A	Sample B
Snails	40	3
Mites	150	30
Spiders	10	40
True worms	10	0
Centipedes	5	1
Insects — Ants	30	5
Springtails	140	65
Aphids	70	100
Midges	110	20
Beetles	50	10

* (Numbers simplified from actual data)

(*a*)   (i) Which animal was present in the largest number at site A?   (*1 mark*)

(ii) Which animal was present in the largest number in the combined samples, A and B?
                                                                                                          (*1 mark*)

(*b*) Complete the circle below to form a pie-chart of the insects at site A. The circle has been divided into 20 equal parts. The sector for the ants has been completed on the pie-chart to help you.   (*3 marks*)

Insects at site A

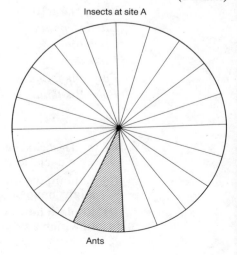

Ants

(c) From the numbers given in the table, which animal is likely to be a secondary consumer?
(*1 mark*)

(d) Suggest *two* reasons why there are more snails in sample A than in sample B. (*2 marks*)
LEAG

66 (a) Explain how you would use pitfall traps to demonstrate that more animals are found around the area of a compost heap than in the middle of a lawn. (*3 marks*)

(b) (i) Name *two* organisms you might find in the traps. (*1 mark*)

(ii) To which group of animals do these organisms belong? (*1 mark*)
MEG

67 (a) Soil from a deciduous wood in the British Isles contains an abundance of microorganisms, annelids and wingless insects.

(i) Name the *two* main groups of microorganism that you would expect to find. (*2 marks*)

(ii) Name *one* annelid and *one* wingless insect that you would expect to find. (*2 marks*)

(b) 10 mm discs cut from fallen oak leaves were placed in nylon bags of various mesh sizes and buried about 3 cm down in the soil in a deciduous wood. The bags were dug up every month, the area of leaf discs remaining was measured and the bags returned to the soil. It was found that the total area of leaf material decreased most rapidly in the bags of 7 mm mesh, about seven times more slowly in the bags of 0.5 mm mesh and that there was hardly any decrease at all in the bags of 0.003 mm mesh even after six months. The experimenters assumed that all leaf material that disappeared was broken down and concluded that the rate of breakdown was determined only by the size of the mesh.

(i) List the organisms named in your answer to (*a*) in the appropriate part of the following table.

Size of mesh	Organisms which could *not* pass through the mesh
7.0 mm	
0.5 mm	
0.003 mm	

(*4 marks*)

(ii) Comment on the assumption and conclusion of the experimenters, stating clearly what criticisms may be made of them. (*4 marks*)

(c) Examination of the faeces of millipedes (small herbivorous arthropods) feeding on fallen oak leaves revealed that their faeces were very similar to the leaf litter except that the oak leaves were now in very small fragments. In view of this finding an investigation was carried out on the rate of carbon dioxide release from entire oak leaf litter, ground up oak leaf litter and millipede faeces. The following results were obtained.

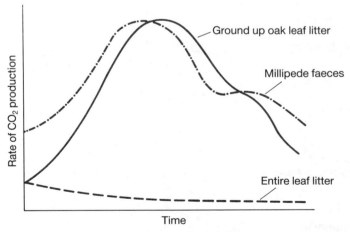

(i) The rate of carbon dioxide release was taken as a measure of the extent of microbial activity. Explain the reasoning behind this. (*3 marks*)

(ii) In view of the information provided in this and in earlier sections of the question, put forward a hypothesis to explain the part played by soil invertebrates in the breakdown of leaf litter. State clearly what support your hypothesis is given by the results described earlier. (*5 marks*)
NISEC

68 The diagram below shows a section through the soil found in an area of forest growing on chalk. The forest contains oak, beech and sweet chestnut trees.

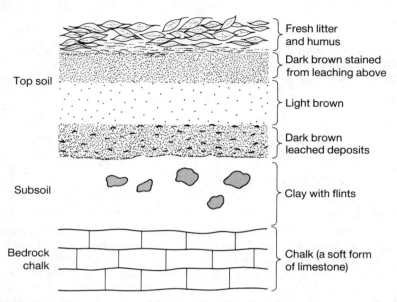

(*a*) There is a deep layer of litter and humus at the surface of the soil.
    (i) What does the litter come from? *(1 mark)*
    (ii) What is humus? *(1 mark)*
    (iii) How is humus formed? *(1 mark)*
    (iv) Eventually the litter becomes a part of the topsoil. Name a group of organisms responsible for this. *(1 mark)*

(*b*) After a period of heavy rain, the soil becomes very waterlogged. Using the information in the diagram above, suggest why this happens. *(1 mark)*

(*c*) The apparatus shown below was used in an experiment to investigate the numbers and variety of organisms found in topsoil and in subsoil.

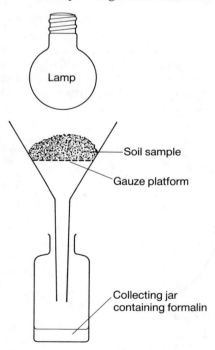

    (i) What is this apparatus called? *(1 mark)*

    (ii) Why is it important to use equal amounts of topsoil and subsoil in the experiment? *(1 mark)*

    (iii) What is the purpose of the lamp? *(2 marks)*

(*d*) More organisms are found in the topsoil than in the subsoil. Give *two* reasons for this. *(2 marks)*

LEAG

69 The diagrams in the table below show four different kinds of woodlouse.

Examine the diagrams carefully, then describe *one* way in which each kind of woodlouse differs from the other three.

Ignore differences in size.

Woodlouse	Feature which is different from the other three
Oniscus asellus	
Porcellio scaber	
Armadillidium vulgare	
Philoscia muscorum	

70 Some of the characteristics of six British arthropod groups are shown in the table below.

Name of group	Number of legs	Number of wings	Appearance of wings
Isopoda	7 pairs	None	None
Diptera	3 pairs	1 pair	Transparent
Coleoptera	3 pairs	2 pairs	Inner wings transparent; outer wings hard and shell-like
Arachnida	4 pairs	None	None
Myriapoda	More than 7 pairs	None	None
Lepidoptera	3 pairs	2 pairs	Both pairs large and scaly

Construct a key, consisting of numbered paired statements, which will enable the arthropod groups to be identified.

SEB

71 (*a*) The diagram represents a transect produced by pupils during a visit to a rocky shore habitat.

Complete the table below the diagram by placing ticks in the appropriate boxes to indicate the presence of the different organisms in the different areas of the shore between low water mark (LWM) and high water mark (HWM). Two ticks have already been placed in the boxes.

(9 marks)

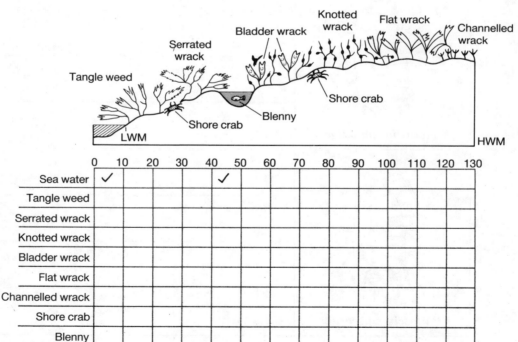

	0	10	20	30	40	50	60	70	80	90	100	110	120	130
Sea water	✓				✓									
Tangle weed														
Serrated wrack														
Knotted wrack														
Bladder wrack														
Flat wrack														
Channelled wrack														
Shore crab														
Blenny														

(*b*) In one area of the rocky shore 50 small crabs (A) were found and marked with a small dot of paint before being released in the same area. One week later 50 crabs (B) were again caught in the same area. It was noticed that only 5 (C) of those caught the second time were marked with the paint. An estimate of the size of the crab population in the area can be obtained by using the following equation:

$$\text{Total population} = \frac{\begin{array}{c}\text{Number of animals} \\ \text{caught and marked} \\ \text{the first time}\end{array} \times \begin{array}{c}\text{Number of animals} \\ \text{caught the second} \\ \text{time}\end{array}}{\begin{array}{c}\text{Number of marked animals caught the second} \\ \text{time.}\end{array}}$$

or total population $= \dfrac{A \times B}{C}$

Use the information provided to estimate the total number of crabs in the population on that part of the shore. Show your calculations clearly. *(2 marks)*

NISEC

Unit 20

72 A digging over the soil
B rotating the crops grown in an area
C adding farmyard manure
D lime spreading
E scattering artificial fertilizer

Which of the above gardening practices:

(*a*) quickly increases the plant nutrient level in the soil?
(*b*) helps control infectious diseases in plants?
(*c*) improves the water retaining properties of a sandy soil?
(*d*) raises the pH value of the soil?

NISEC

73 For each group of three terms write a short paragraph in which you demonstrate how the terms are biologically linked.

(*a*) Exercise, carbon dioxide, ventilation rate. *(4 marks)*
(*b*) Fertilizer run-off, algal growth, fish death. *(4 marks)*

(Unit 9 may also be helpful if guidance is needed whilst answering this question) *NISEC*

74 The table below records the number of deaths from diphtheria in Northern Ireland between 1934 and 1958.

Years	1934	1938	1942	1946	1950	1954	1958
Number of deaths per year	114	100	90	28	4	1	0

On the axes below choose a suitable scale for the death rate and construct a line graph of the information provided in the table. *(5 marks)*

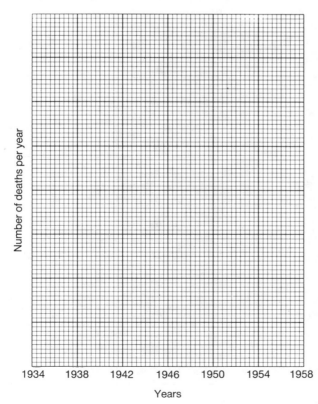

Use your graph to estimate:

(*a*) (i) the year in which 94 people died of diphtheria; *(1 mark)*

(ii) which *two year period* had the greatest drop in the number of deaths. *(1 mark)*

(*b*) Babies and young children are now immunized against diphtheria and have been since 1900. Describe fully how immunization protects the child against the disease. *(4 marks)*

(*c*) Since 1945 the number of babies successfully immunized has risen steadily. How does the graph reflect this increase in the number of babies immunized? *(2 marks)*

NISEC

Unit 21

75 The diagram shows a student heating a liquid in a Bunsen flame.

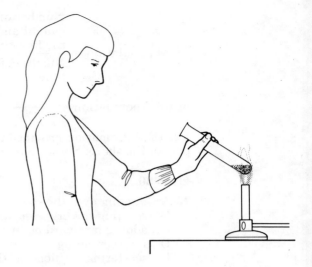

List *three* mistakes the student is making. *(3 marks)*
NEA

76 A group of pupils collected pond snails from the school pond. The shells were measured accurately and the results are shown opposite.

Sizes of snails in mm

5	14	17	9	21	19	25	27	33	34	30
18	6	7	16	11	22	23	23	26	29	31
8	19	9	18	15	13	17	29	22	27	25
12	16	10	11	21	24	25	28	19	21	24
19	25	20	23	23	28	20	20	10	15	23
21	28	22	26	21	27	30	18	7	8	20
26	24	26	31	30	24	32	22	24	10	17

(*a*) Use the data to complete the table opposite.

Size of snail shell in mm	Number of snail shells of this size
5–6	
7–8	
9–10	
11–12	
13–14	
15–16	
17–18	
19–20	
21–22	
23–24	
25–26	
27–28	
29–30	
31–32	
33–34	

(3 marks)

(*b*) Use the results obtained in the table to plot a bar graph of the number of snails against shell size.

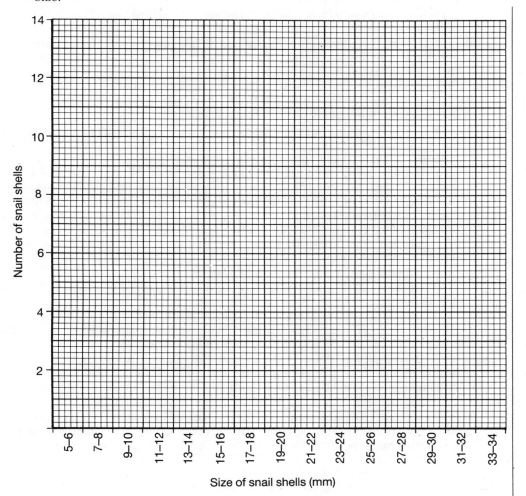

(*c*) Suggest *one* explanation for the results shown in your bar graph. (*1 mark*)
LEAG

77 Scientists studying the colonization of land which had been used for the testing of nuclear bombs noticed that very few of the seeds produced by the colonizing plants germinated.

You have ordinary laboratory apparatus and some seeds. You also have a box containing a weak radioactive source into which you can *safely* place your seeds.

Describe how you would carry out an experiment to find out if radioactivity affects the numbers of seeds which germinate. (*5 marks*)
NEA

Answers to Specimen Questions

Key: /......./ = 1 mark; /....../ / = 2 marks; /....../// = 3 marks.

1(a)(i) cell wall (ii) chloroplast (iii) cell membrane
(b)(i) cell wall (ii) large vacuole — present in plant cells but *not* in animal cell; (chloroplast would also score)
(c) **A** — on outside of root   **B** — palisade layer of leaf   **C** — inside blood vessels
(d) **A** — has a large surface area   **B** — large number of chloroplasts
(e) (i) any one of: water/a named mineral salt *or* ion (ii) oxygen
2 (a) nucleus (b) cell wall (c) (i) cell membrane (ii) oxygen, water
3 (a) **P** — fish   **Q** — mammal   **R** — bird
(b) **P** — scales/fins   **Q** — any two of: hair/(external) ears/mammary glands   **R** — wings, feathers
4 D
5 E
6 (a) **A** (b) **E** (c) **A** (d) **B**

7 (a) **A** (b) temperature too low (in E) (c) bacteria could enter C, but were excluded by the bung in B (d) control tubes (for comparison with the other tubes in which live bacteria were present) (e) prevents milk from gaining bacteria from the test tubes (f) unequal quantities of milk used

8 (a) although Streptomycin appears best it is ineffective against some colonies. Terramycin and Erythromycin seem equally effective
(b) it is unaffected by Sulphafurazole and Chloramphenicol but sensitive to the rest
(c) bacteria are killed by some antibiotics and not others
(d) subculture the 3 colonies on fresh agar. Add a Streptomycin disc. If no clear zone appears, bacteria are resistant to Streptomycin

9 (a) bacteria *or* fungi/on skin/enter tomato at a break/secrete enzymes to feed/grow into a colony or mycelium/digesting the tomato/cause waste products which smell or taste bad/moisture level (inside bag) encourages decay organisms to survive (any five give 5 marks)
(b) skin damage/humid conditions/going bad when bought/skin wet against polythene (of some tomatoes, all help these ones to go bad) — (any two give 2 marks).
(c) large number of tomatoes tested, e.g. 10/same number as controls/only one difference in conditions between test and control/apparatus sterilized before use/temperature specified/duration of experiment stated/apparatus described (? diagram)/method of getting result, e.g. 7/10 went bad (not 'most went bad')/ method of recording result, e.g. table (any eight good points give 8 marks)

10 (a) (i) women age 25–29 (ii) 20–24 (iii) 30–39
(b) (i) fat (or lipid) (ii) under skin *or* around kidney etc. (iii) e.g. heart disease *or* difficulty with breathing
(c) (i) respiration/results in loss of carbon dioxide/and water/from all cells/by their metabolism (any three give 3 marks) (ii) increased respiration/especially in muscles/owing to extra energy needed. Loss of sweat (any two give 2 marks)
(d) (i) fibre helps peristalsis (ii) provides vitamin C/minerals/fibre (any one) (iii) not fattening/avoids heart disease (any one) (iv) not fattening/avoids tooth decay (any one)
NB You are unlikely to get a mark for the same answer in two places, e.g. 'not fattening' in both (iii) and (iv)

11 (a) (i) 4060 kJ (ii) gains weight/gets fat (one) (iii) physically active/pregnant (one)
(b) (i) breakfast (i.e. 70:14) (ii) 2:1 (iii) high cholesterol level in blood/'furred up' arteries (i.e. atherosclerosis)/reduced O_2 supply/coronary artery blockage/heart attack (any two)
(c) chocolate/2.4 mg per g (i.e. $120 \div 50$)
(d) iron/anaemia *or* vitamin C/scurvy
(e) excessive protein intake (which is deaminated)

12 (a) (i) to exclude light (ii) central black diamond/areas outside foil, black (iii) starch
(b) otherwise light reaches one side/helps fix stencil to leaf (one)
(c) (i) chlorophyll (ii) ethanol catches fire easily

13 (a) (i) two correct labels (see Fig. 5.6) (ii) on 6 palisade cells/10 spongy mesophyll cells/2 guard cells
(b) (i) for respiration (release energy)/to make cellulose/for storage as starch (one) (ii) energy food/ sweetener/preservative (one)
(c) (i) dull/cloudy/ less sun (one) (ii) a line at level of highest peak, marked X

14 (a) (i) water (ii) oxygen
(b) absorbs or traps or converts light energy
(c) provides food/and oxygen/ removes CO_2 (any two)
(d) (i) kills leaf/stops chemical activity in leaf (one) (ii) chlorophyll is removed (iii) leaf turns blue-black colour

15 (a) **A** — enamel **B** — dentine **C** — pulp
(b) (i) acids produced/these dissolve calcium salts (*or* enamel) (ii) pain/tooth may have to be extracted/stops bacteria spreading (any two)
(c) (i) hardens enamel (ii) neutralizes acids

16 (a) Liver: bile salts/emulsify fats/to droplets. Pancreas: amylase/turns starch to sugar (maltose)/lipase/turns fats to fatty acids and glycerol/protease/turns protein to amino-acids./These enzymes use water to digest complex molecules (*or* hydrolyse them)
(b) (i) glucose in blood/enters liver cells/becomes glycogen/insulin from pancreas promotes this (ii) liver: excess *amino acids* /are broken down (deaminated) to urea/for exretion by kidneys/worn out *red blood cells* broken down giving bile pigments/for excretion in faeces

17 (a) (i) dentine (ii) correct label (iii) enamel
(b) wild horses have coarser diet/(wearing down enamel)/of continuously growing teeth (any two)
(c) smaller/not continuously growing/no exposed dentine/roots not open (any two)

18 (a) 0.42 cm^3 (b) 0.7 cm^3 (c) (i) wilted (ii) water loss exceeded uptake *or* cells no longer turgid owing to water loss
(d) line drawn in top left of graph, same origin

(e) diagram of a suitable potometer sufficiently labelled

19 (a) (i) *diffusion*: movement of molecules (*or* gas *or* solute *or* ions)/from region of high concentration/to region of low concentration (ii) *osmosis*: passage of water (solvent)/across a selectively permeable membrane/from a less concentrated to a more concentrated solution (iii) *active uptake*: movement of solute molecules (or ions)/requiring energy from the cell/across a cell membrane

(b) (i) water by osmosis/by diffusion through cell wall; ions *or* mineral salts/by active transport (3) (ii) CO_2 in/and O_2 out (during photosynthesis) by diffusion. O_2 in/and CO_2 out (during respiration) by diffusion. Water in by veins/and out by transpiration (4) (iii) glucose *or* amino-acids in/through capillaries. Fatty acids and glycerol in/through lacteals (4)

20 (a) (i) **A** (ii) stomata on lower surface/vaseline seals them reducing water loss

(b) leaf has lost enough water to cause stomata to close

(c) 0.037–0.039 g (sum of losses in A and B, ÷ 2)

(d) 1. dry habitat 2. shows little loss of water/so must be adapted to a habitat where water is short

21 A

22 (a) to remove any starch that might be on the outside

(b) same as body temperature/optimum temperature for enzyme (one)

(c) digests it *or* makes it into smaller molecules/produces reducing sugar/

(d) add Benedict's solution and warm/look for colour change from blue to green *or* yellow *or* orange

(e) boiling destroys enzyme *or* prevents action of saliva

(f) selectively permeable/small molecules, only, pass through/sugar molecules, not starch, pass through (any two)

(g) Visking tubing is non-living/no capillaries/no villi *or* large inner surface area/no movements take place (any two)

(h) (i) diffusion is slower at lower temperature (ii) enzyme action is slower so sugar production is slower

23 (a) **C** — oxygen **G** — any named product of digestion/water/salts/vitamins (one) **H** — water

(b) carbon dioxide/water (one)

(c) **B** — breathing in/inspiration (one) **D** — peristalsis/ swallowing/vomiting (one) **F** — churning/mixing (one)

(d) **D**, e.g. drugs, food additives/and their harmful effects **E**, e.g. carbon monoxide, nicotine, SO_2, asbestos (any two)/and their harmful effects

24 E

25 (a) **X** — right ventricle **Y** — left atrium (auricle)

(b) **A** — vena cava **B** — pulmonary vein

(c) A had less O_2 more CO_2 than B

(d) Z pumps blood all round body/X only to lungs

(e) (i) coronary arteries (ii) food/O_2 supply to muscle is stopped/cardiac muscle/ceases to work well

26 (a) **D** — right atrium **E** — right lung

(b) **I** (c) (i) blocking/fat deposits/narrower bore for blood (one) (ii) heart attack/reduced blood supply/less 0_2/less food (one)

27 (a) sprints — faster/long distance races — slower

(b) 2 suitable examples, e.g. at 100 m and 10000 m

28 (a) carbon dioxide

(b) lime water *or* bicarbonate indicator solution

(c) increasing rate from 25°C to 40–50°C/faster decline in rate above this

(d) increase because reaction rates increase with temperature rise/until higher temperatures destroy enzymes

(e) (i) aerobic = 38 ATP, anaerobic = 2ATP *or* aerobic ATP production is 19 × anaerobic ATP production (ii) under anaerobic conditions there is a decrease in rate of reproduction/*or* number of yeast cells *or* growth of colony/slower uptake of nutrients/growth curve shifts to right

(f) (i) no chlorophyll/chitin cell wall or cell walls similar/never store starch/do not move (any two) (ii) yeast is unicellular — not a mycelium

29 (a) energy content of peanut

(b) temperature rise is required/subtract initial from final temperature

(c) heat lost to air/and to test tube/soot *or* remains of peanut show that peanut does not burn fully (any two)

30 (a) half mark for each term used correctly, e.g. 'if sugar is respired without oxygen it gives a little energy and often carbon dioxide as a by-product.'

(b) (i) becomes partly anaerobic (ii) lactic acid

31 Any five from: body cooler at night and warmer by day/greater variation in hot climate *or* less variation in cold climate/coldest around 0300 hrs/hottest in early afternoon/similar variation in both climates/daily variation averages about 1°C/body temperature not constant etc

32 (a) 1 — cortex 2 — medulla *or* pyramid 3 — ureter
(b) (i) 10% (ii) 1 (iii) reabsorbed/from tubules or nephrons/into blood (any two)
(c) (i) no sweating/blood too dilute/excess water is removed (ii) intake of salt the same/both salt concentration/and water (any two) are regulated by kidney/to maintain homeostasis (any two) (iii) to eliminate wastes *or* urea/excretion/remove toxins (any two) (iv) drink/eat salt

33 Any 14 points from: urea dissolved in blood/of hepatic vein/vena cava→/heart→/pulmonary (lung) circulation→/heart→/aorta→/renal artery→/capillaries *or* glomerulus/of nephron/filtered/into (Bowman's) capsule/by pressure (ultra filtration)/in solution/down tubule/detail of tubule structure/not reabsorbed into blood/collecting duct/pelvis/down ureter in urine/by peristalsis/storage in bladder/diagram giving relevant information.

34 (a) **A** — hair **B** — epidermis **C** — erector muscle
(b) (i) sweat produced/from sweat gland/along duct/evaporates using the excess heat/cools skin/and blood/which by circulation cools the body (any four) (ii) heat carried by blood/goes through surface capillaries/by vasodilation/or arteries/heat lost by radiation/by conduction/cools blood/which circulates (any four)
(c) secrete oil (grease) which waterproofs/reduces evaporation/keeps skin supple — no cracks/kills bacteria/prevents infection (any three)

35 (a) correct drawing and labelling of four structures (see Fig. 11.2)
(b) **B** at blind spot
(c) reduced field of vision/distance estimation poorer (no binocular vision)
(d) ciliary muscles relax/ligaments pull (tighten)/to make lens thinner
(e) (i) pupil wider (ii) radial muscles contract/relax the circular muscles (iii) larger pupil/allows ultra-violet light in

36 **E**

37 **E**

38 (a) side shoots grow/lateral (axillary) buds sprout (any one)
(b) see the effect of lanolin paste/control (any one)
(c) controls (suppresses) lateral bud development

39 (a) **E** (b) **B**

40 Any three points from:
strength of solid bone/arranged into cylinder/light weight/cross struts/giving support in different directions; *cartilage* at joint/smooth movement/spongy head/giving lightness; *processes*/for muscle attachment *ends broad*/ to spread weight

41 (a) **D**
(b) **C**
(c) **B**

42 (a) **A** — anther *or* stamen **B** — stigma **C** — style **D** — sepal **E** — ovule **F** — ovary
(b) receive pollen
(c) fruit
(d) large petals/small stigma/position of stamens and stigma (any two)
(e) remove immature stamens (anthers)
(f) leaves not present to obstruct air-borne pollen/greater likelihood of pollination

43 1 A 2 D 3 E 4 B 5 C

44 (a) ovule
(b) (i) correct proportion of seed(s) for a test tube/wet cotton wool in contact with seeds (ii) soak seeds/minimum 5 seeds per test tube/same number in all test tubes/minimum 3 different temperatures/temperatures specified/how kept constant/other conditions the same/wet cotton wool (if not mentioned in (b)/what 'germination' meant, e.g. split testa *or* radicle emerged/what was timed, e.g. first seed to germinate *or* 50% of them germinated *or* number germinated after a sensible period of time (any seven) (iii) any sensible difficulty, e.g. keeping temperature constant, *or* keeping seeds equally moist

45 (a) (i) prevents decay/stimulates growth (one) (ii) more air (iii) contact with water/prevents rocking which may break rootlets (one)
(b) (i) reduces transpiration *or* dehydration/less wind resistance (which rocks stem) (ii) no photosynthesis/stem cannot get food/bacteria may enter (any two)
(c) (i) use rooting powder/hormones/use fungicide/enclose under cloche (any one) (ii) explanation, e.g. hormones improve rooting/fungicide reduces infection (any one)
(d) (i) several cuttings of both plants/in same conditions/described/measure heights/and count side branches/take averages (any four) (ii) different heights/or side branches in the two sets of cuttings (one) (iii) no difference between the two sets of cuttings.

46 (a) **A** — placenta **B** — umbilical cord **C** — amnion **D** — cervix
(b) diffusion/through placenta

(c) protection/support (one)

(d) contract/forcing baby through cervix

(e) (i) mutation/extra chromosome present in ovum (one) (ii) one extra chromosome/47 chromosomes, not normal 46 (one)

47 (a) (i) ovary/secretes hormone into blood stream/no duct (ii) testosterone/testis

(b) (i) suckling causes high prolactin levels//which may cause low oestrogen levels//thus no ovulation/ — no eggs to fertilize (ii) high prolactin linked to low oestrogen//no ovulation/ and so no fertilization

48 (a) **A** — 52.5/53.0 mm **B** — 47.0/47.5 mm

(b) **A** — 185 mm **B** — 205 mm

(c) day 14–day 16

(d) 180–112 = 68 /68 ÷ 2 = 34 mm per day

(e) differences in health *or* disease/in food content of seeds/soil *or* humus content/soil microorganisms (any three)

49 (a) water necessary/other variables, e.g. light, not essential

(b) contents *either* alkaline pyrogallol/no oxygen/no aerobic respiration/so no energy for growth; *or* potassium hydroxide/no carbon dioxide/no photosynthesis/so no food for growth; *or* toxic substance/killed seedling/no respiration *or* photosynthesis/no growth possible

(c) (i) **B** normal growth/chlorophyll normal **E** rapid spindly growth/leaves lack chlorophyll (ii) lower metabolic rate *or* enzyme activity in **D** than **E**/(temp)/**D** lacks chlorophyll, as in **E**/(no light for both)

(d) **G** had germinated under good conditions/but something, e.g. disease, from soil, killed seedlings/**H** had no water to germinate with

(e) **E** had more water in its cells/no photosynthesis to gain mass/only respiration to lose mass. **B** gained dry mass by photosynthesis

(f) e.g. X causes stunting *or* short stem growth/withstands wind better so plants not blown over to spoil grain *or* trace elements/give better growth

50 VZUXWY

51 (a) nucleus

(b) controls inherited features (characteristics)/controls manufacture of enzymes (for metabolism)/duplicated exactly at mitosis (to pass on 'recipe' for life of cell) (any two)

(c) (i) one form of a gene (of a number of alternatives for a characteristic) (ii) e.g. tongue-rolling

52 (a) e.g. grid with ticks   (b)

	Long	Short
Male	29	21
Female	26	24

(c) Ll and ll (where L = long; l = short)

53 (a) 46 (body) 46/23 (gametes) 23/46 (zygote)/46 (twins) 46

(b) (i)

1	tall	short	($\frac{1}{2}$)
2	round	wrinkled	($\frac{1}{2}$)
3	yellow	green	($\frac{1}{2}$)
4	grey	white	($\frac{1}{2}$)

(ii) the only characteristic which appears in the F1 generation (2)

(c) (i) 705 ÷ 244/ = 2.89:1 (ii) 3:1

(d) (i) TT × tt (ii) Tt (iii) Tt   ×   Tt

T t   T t

TT Tt   tT tt

54 (a) accurate *straight* lines joining points/deduct one mark for each point wrongly plotted (to max. 3 marks)

(b) 75 per 10 000 (74–76 allowed)

(c) 16.7 (100 ÷ 6 = 16.66)

(d) any two of: mutation more likely/longer time for same chance/ageing egg cells/more time to be exposed to radiation *or* other causes of mutation.

55 (a) blood does not clot/when blood vessel is injured (b) 'XY' means male//heterozygous for sex chromosomes//'h' means recessive gene//

(c) wife of X^hY male is most likely $X^H X^H$, giving 50% chance of X^HX^h daughter///husband of daughter most likely X^HY giving 25% chance of X^hY son///*or* 12.5% chance of haemophiliac grandchild (any six points)

56 (a) **A** — dark/not nicked or tattered *or* **B** — pale (peppered)/damaged (any two)

(b) (i) to identify the recaptured moths/mark not noticeable (e.g. to predators) when they rested (ii) dark one (iii) dark ones less visible (better camouflaged) on darkened trees/predators take fewer of them *or* light ones more obvious/predators take more of them

57 (a) artificial

(b) many different genes (alleles) to choose from

(c) (i) asexually/vegetatively/by cloning, e.g. tissue culture (ii) sexually reproduce/detail of how/to introduce other useful genes (alleles)/induce mutations use genetic engineering (any two)

58 (a) less likely to get malaria
(b) sickle cells/malaria will often kill people with normal cells

59 **A** — nitrogen fixation **C** — absorption/active transport (one) **E**— feeding **F** — excretion/egestion/death (one) **G** — decay

60 (a) (i) rabbit/bird/insect (one) (ii) any one example from 'foxes' to 'grass snakes' column (iii) same as (ii)
(b) recycling nutrients//turning organic materials into inorganic//(either)
(c) small dose of insecticide on each insect eaten/builds up into a bigger dose in each bird
(d) e.g. reduced nesting sites for bird prey/reduced protection for other food, e.g. mammals/less food for prey animals (any three)

61 (a) (i) old exoskeleton split off/soft new exoskeleton beneath enlarges/because old skeleton restricts growth *or* does not grow with the body (ii) any two arthropod or insect features (see unit 21.7)
(b) food assimilated = leaves consumed − faeces produced; food respired = food assimilated − insect mass
(c) $5 \rightarrow 15$ mg in 3 growth stages/$15 \rightarrow 45$ mg in two growth stages/therefore faster growth later//
(d) (i) privet = producer/stick-insect = primary consumer (ii) $\frac{45}{500} \times \frac{100}{1} / = 9\%$ (iii) cuts out a trophic level/reducing losses in food/therefore less plant food needed to give same food value//

62 (a) (i) water boatmen (ii) mites
(b) (method: number of insects = $30 + 170 + 50 + 120 + 30 = 400$; therefore damsel-fly nymphs = $\frac{30}{400} \times 100$ per cent = 7.5%
Each segment of pie chart is $\frac{100}{20} = 5\%$; therefore pie slice for damsel-fly nymphs is 1.5 segments) one mark for each of three correctly drawn and labelled segments
(c) leech
(d) food is there/snails do not float in mid-pond/better shelter/fewer predators etc. (any two)

63 *Bdellocephala* = **E**/*Crenobia* = **A**/*Dugesia* = **C**
Polycelis felina = **D**/*Polycelis tenuis* = **B**

64 (a) apparatus/how used (2 marks). Number of times sampled *or* large number collected/counting/different areas *or* levels/how samples were made comparable (any three)
(b) humans: felling trees *or* coppicing/collecting firewood/picking flowers/trampling/hunting/crop spraying/stubble burning etc. (max. four); other animals: predation/pollination/grazing/faeces/death/dispersal etc. (max. four)
(c) sulphur oxides, e.g. SO_2/nitrogen oxides, e.g. NO_2/carbon dioxide/dissolved in water *or* rain/kills lichen/decreases soil pH/decreases litter pH/reduces decomposition/may cause leaf fall etc. (max. three). Reduce combustion/fit filters on chimneys/filters on car exhausts/legislation etc (max. two)

65 (a) (i) mites (ii) springtails
(b) (see **Q62** for method) one mark for each of three correctly drawn and labelled segments.
(c) centipedes
(d) more food/better shelter/fewer predators etc. (any two)

66 (a) same number of traps in the two areas/set at same time and collected at same time/number trapped, recorded.
(b) (i) any two sensible examples (one mark)
(ii) correct group e.g. earthworm = annelid, for each example (one mark)

67 (a) (i) bacteria/fungi (ii) annelid, e.g. earthworm/wingless insect, e.g. springtail
(b) (i) none (7 mm)/0.5 mm earthworms/0.003 mm springtails (collembola) (ii) leaf discs or fragments could have been removed// from larger mesh bags by invertebrates and deposited elsewhere//
(c) (i) microbial activity (decay)/gives off CO_2/from respiration (ii) invertebrates chew litter finely/giving large surface area/for microorganisms to attach/gut microbes may assist/more rapid breakdown.

68 (a) (i) fallen leaves (ii) litter incorporated into soil/organic material formed from litter (one) (iii) activities of microorganisms/fungi/bacteria (iv) earthworms
(b) clay layer — relatively impervious to water
(c) (i) Tullgren funnel (ii) allow comparison of numbers of organisms in equal volumes of soil (iii) animals move away from heat/or light/or dry soil (any two)
(d) more food available/better soil atmosphere etc. (max. two)

69 *Oniscus:* only 6 pairs of legs visible/*Porcellio*: spotted/*Armadillidium*: no cerci (pair of 'tails')/*Philoscia*: leg arrangements.

70 (see unit 19.8 for method)

71 (a)

	0	10	20	30	40	50	60	70	80	90	100	110	120	130
Sea water	√				√									
Tangle weed		√												
Serrated wrack			√	√										
Knotted wrack						√	√	√	√	√				
Bladder wrack						√	√							
Flat wrack										√	√	√		
Channelled wrack												√	√	
Shore crab			√					√						
Blenny					√									

(one mark for each horizontal line correct)

(b) $\dfrac{50 \times 50}{5} = 500$ (two marks)

72 (a) **E** (b) **B** (c) **C** (d) **D**

73 (a) exercise requires increased metabolism *or* respiration/resulting in increased CO_2 production/extra CO_2 sensed by respiratory centre in brain/nerve impulses cause increase in breathing (ventilation) rate.

(b) fertilizer washed into fresh water by rain/causes increased algal growth (by eutrophication)/their death — decomposition by bacteria/reduces O_2 level in water — so fish die.

74 Graph: suitable scale (1) accurate plotting of points (3) points joined by *straight* lines (1) (see unit 22.7)

(a) (i) 1940 (ii) 1944–1946

(b) attenuated (weakened) bacteria introduced to body/are antigens/to which body reacts by producing antibodies/enough antibodies give body immunity.

(c) increased immunization → fall in deaths/46 → 0

75 No goggles (safety glasses)/test tube pointing towards student/no test-tube holder/test tube in wrong part of flame (any three)

76 (a) 1 mark deducted for every 2 errors (max. 3)

(b) plots on graph must match *own table*: 1 mark deducted for every 2 errors (max. 3)

(c) snails of different age/or species/or amount of food eaten etc. (one)

77 batches of seeds/damp, e.g. cotton wool to grow on/warmth/one lot exposed to radioactive source and one not/count number affected

Page numbers in bold type indicate the main references for those entries.